Principles of Protein–Protein Association

About the Series

The Biophysical Society and IOP Publishing have forged a new publishing partnership in biophysics, bringing the world-leading expertise and domain knowledge of the Biophysical Society into the rapidly developing IOP ebooks program.

The program publishes textbooks, monographs, reviews, and handbooks covering all areas of biophysics research, applications, education, methods, computational tools, and techniques. Subjects of the collection will include: bioenergetics; bioengineering; biological fluorescence; biopolymers *in vivo*; cryo-electron microscopy; exocytosis and endocytosis; intrinsically disordered proteins; mechanobiology; membrane biophysics; membrane structure and assembly; molecular biophysics; motility and cytoskeleton; nanoscale biophysics; and permeation and transport.

Principles of Protein–Protein Association

Harold P Erickson

Department of Cell Biology, Duke University, Durham, North Carolina, USA

IOP Publishing, Bristol, UK

ISBN   978-0-7503-2412-0 (ebook)
ISBN   978-0-7503-2410-6 (print)
ISBN   978-0-7503-2411-3 (mobi)

DOI   10.1088/2053-2563/ab19ba

Version: 20190601

IOP Expanding Physics
ISSN 2053-2563 (online)
ISSN 2054-7315 (print)

British Library Cataloguing-in-Publication Data: A catalogue record for this book is available from the British Library.

Published by IOP Publishing, wholly owned by The Institute of Physics, London

IOP Publishing, Temple Circus, Temple Way, Bristol, BS1 6HG, UK

US Office: IOP Publishing, Inc., 190 North Independence Mall West, Suite 601, Philadelphia, PA 19106, USA

Contents

Preface

This volume originated from lectures I have been giving to graduate students. The students are mostly first and second year graduate students from the Duke University Program in Cell and Molecular Biology. I presume only a basic knowledge of biochemistry. I highly recommend that students review basic principles of protein structure prior to the course. Excellent sources are the texts: *Molecular Biology of the Cell*, by Alberts *et al*, chapter 3, 'Proteins'; or *Cell Biology* by Pollard and Earnshaw, chapter 2 'Molecular Structures.'

I also highly recommend that students download a protein structure viewer and use it to image on their own computer the structures displayed in the figures. Recommended viewers are Chimera, Pymol and KING.

In recent years the course has comprised six sessions of 80 minutes each, where I present background material and then lead discussion of the assigned papers. The chapters included here have evolved from my notes for these class sessions. These notes may be useful for faculty organizing similar classes, and/or for self-instruction of students and researchers who find a need to understand principles of protein–protein association.

Author biography

Harold P Erickson

Harold P Erickson received his PhD in Biophysics from Johns Hopkins University and did postdoctoral work in Cambridge, England. He joined the faculty at Duke University Medical Center in 1971, and is currently James B. Duke Professor in the Departments of Cell Biology, Biochemistry and Biomedical Engineering. His research has spanned two broad areas of cell biology: cytoskeleton (microtubules and the bacterial tubulin homolog, FtsZ); and extracellular matrix (fibrinogen, tenascin and fibronectin). He has contributed to several areas of electron microscopy (image processing, negative stain and rotary shadowing) and has theoretical work on the thermodynamics of cooperative assembly and diffusion-limited kinetics of protein–protein association.

IOP Publishing

Principles of Protein–Protein Association

Harold P Erickson

Chapter 1

Size and shape of protein molecules at the nm level determined by sedimentation, gel filtration and electron microscopy

This chapter was published in 2009 *(Biol. Proced. Online* 11:32–51*). It presents methods and calculations that are fundamentally important to determining the size and stoichiometry of protein complexes. It is reprinted here to have these resources readily available.*

An important part of characterizing any protein molecule is to determine its size and shape. Sedimentation and gel filtration are hydrodynamic techniques that can be used for this medium resolution structural analysis. This review collects a number of simple calculations that are useful for thinking about protein structure at the nm level. Readers are reminded that the Perrin equation is generally not a valid approach to determine the shape of proteins. Instead, a simple guideline is presented, based on the measured sedimentation coefficient and a calculated maximum S, to estimate if a protein is globular or elongated. It is recalled that a gel filtration column fractionates proteins on the basis of their Stokes radius, not molecular weight. The molecular weight can be determined by combining gradient sedimentation and gel filtration, techniques available in most biochemistry laboratories, as originally proposed by Siegel and Monte. Finally, rotary shadowing and negative stain electron microscopy are powerful techniques for resolving the size and shape of single protein molecules and complexes at the nm level. A combination of hydrodynamics and electron microscopy is especially powerful.

1.1 Introduction

Most proteins fold into globular domains. Protein folding is driven largely by the hydrophobic effect, which seeks to minimize contact of the polypeptide with solvent. Most proteins fold into globular domains, which have a minimal surface area.

Peptides from 10–30 kDa typically fold into a single domain. Peptides larger than 50 kDa typically form two or more domains that are independently folded. However, some proteins are highly elongated, either as a string of small globular domains, or stabilized by specialized structures such as coiled coils or the collagen triple helix. The ultimate structural understanding of a protein comes from an atomic-level structure obtained by x-ray crystallography or NMR. However, structural information at the nm level is frequently invaluable. Hydrodynamics, in particular sedimentation and gel filtration, can provide this structural information, and it becomes even more powerful when combined with electron microscopy (EM).

One guiding principle enormously simplifies the analysis of protein structure. The interior of protein subunits and domains consists of closely packed atoms [1]. There are no substantial holes, and almost no water molecules in the protein interior. As a consequence of this, proteins are rigid structures, with a Young's modulus similar to that of Plexiglas [2]. Engineers sometimes categorize biology as the science of 'soft wet materials.' This is true of some hydrated gels, but proteins are better thought of as hard dry plastic. It is obviously important for all of biology, to have a rigid material with which to construct the machinery of life. A second consequence of the close-packed interior of proteins is that all proteins have approximately the same density, about 1.37 g cm^{-3}. For most of the following we will use the partial specific volume, v_2, which is the reciprocal of the density. v_2 varies from 0.70 to 0.76 for different proteins, and there is a literature on calculating or determining the value experimentally. For the present discussion we will ignore these variations and assume the average $v_2 = 0.73$ cm^3 g^{-1}.

1.2 How big is a protein molecule?

Assuming this partial specific volume ($v_2 = 0.73$ cm^3 g^{-1}), we can calculate the volume occupied by a protein of mass M in Da as follows.

$$V(\text{nm}^3) = \frac{(0.73 \text{ cm}^3 \text{ g}^{-1}) \times (10^{21} \text{ nm}^3 \text{ cm}^{-3})}{6.023 \times 10^{23} \text{ Da g}^{-1}} \times M(\text{Da})$$

$$= 1.212 \times 10^{-3}(\text{nm}^3 \text{ Da}^{-1}) \times M(\text{Da})$$

(1.1)

The inverse relationship is also frequently useful: M (Da) $= 825$ V (nm^3).

What we really want is a physically intuitive parameter for the size of the protein. If we assume the protein has the simplest shape, a sphere, we can calculate its radius. We will refer to this as R_{min}, because it is the minimal radius of a sphere that could contain the given mass of protein

$$R_{\text{min}} = (3V/4\pi)^{1/3} = 0.066M^{1/3} \text{ (For } M \text{ in Da, } R_{\text{min}} \text{ in nm)}$$

(1.2)

Some useful examples for proteins from 5000 to 500 000 Da are given in table 1.1.

It is important to emphasize that this is the minimum radius of a smooth sphere that could contain the given mass of protein. Since proteins have an irregular surface, even ones that are approximately spherical will have an average radius larger than the minimum.

Table 1.1. R_{min} for proteins of different mass.

Protein M (kDa)	5	10	20	50	100	200	500
R_{min} (nm)	1.1	1.42	1.78	2.4	3.05	3.84	5.21

Table 1.2. Distance between molecules as function of concentration.

Concentration	1 M	1 mM	1 μM	1 nM
Distance between molecules (nm)	1.18	11.8	118	1180

1.3 How far apart are molecules in solution?

It is frequently useful to know the average volume occupied by each molecule, or more directly, the average distance separating molecules in solution. This is a simple calculation based only on the molar concentration.

In a 1 M solution there are 6×10^{23} molecules/liter, $= 0.6$ molecules/nm^3, or inverting, the volume per molecule is $V = 1.66$ nm^3/molecule at 1 M. For a concentration C, the volume per molecule is $V = 1.66/C$.

We will take the cube root of the volume per molecule as an indication of the average separation.

$$d = V^{1/3} = 1.18/C^{1/3}, \tag{1.3}$$

where C is in molar, and d is in nm. Table 1.2 gives some typical values.

Two interesting examples are hemoglobin and fibrinogen. Hemoglobin is 330 mg ml^{-1} in erythrocytes, making its concentration 0.005 M. The average separation of molecules (center to center) is 6.9 nm. The diameter of a single hemoglobin molecule is about 5 nm. These molecules are very concentrated, near the highest physiological concentration of any protein (the crystallins in lens cells can be at >50% protein by weight).

Fibrinogen is a large, rod-shaped molecule that forms a fibrin blood clot when activated. It circulates in plasma at a concentration of around 2.5 g l^{-1}, about 9 μM. The fibringogen molecules are therefore about 60 nm apart, comparable to the 46 nm length of the rod-shaped molecule.

1.4 The sedimentation coefficient and frictional ratio. Is the protein globular or elongated?

Biochemists have long attempted to deduce the shape of a protein molecule from hydrodynamic parameters. There are two major hydrodynamic methods that are used to study protein molecules—sedimentation and diffusion (or gel filtration, which is the equivalent of measuring the diffusion coefficient).

The sedimentation coefficient, S, can be determined in an analytical ultracentrifuge. This was a standard part of the characterization of proteins in the 1940s and

1950s, and values of $S_{20,w}$ (sedimentation coefficient standardized to 20 °C in water) are collected in references such as the *CRC Handbook of Biochemistry* [3]. Today S is more frequently determined by zone sedimentation in a sucrose or glycerol gradient, by comparison to standard proteins of known S. 5%–20% sucrose gradients have been most frequently used, but we prefer 15%–40% glycerol gradients in 0.2 M ammonium bicarbonate, because this is the buffer used for rotary shadowing EM (section 1.8). The protein of interest is sedimented in one bucket of the swinging bucket rotor, and protein standards of known S (table 1.5) are sedimented in a separate (or sometimes the same) gradient. Following sedimentation, the gradient is eluted into fractions and each fraction is analyzed by SDS–PAGE to locate the standards and the test protein. Figure 1.1 shows an example determining the sedimentation coefficient of BsSMC (the SMC protein from *Bacillus subtilis*).

The sedimentation coefficient of a protein is a measure of how fast it moves through the gradient. Increasing the mass of the protein will increase its sedimentation, while increasing its size or asymmetry will decrease its sedimentation. The relationship of S to size and shape of the protein is given by the Svedberg formula:

$$S = M(1 - v_2\rho)/N_o f = M(1 - v_2\rho)/ (N_o 6\pi\eta R_s) \qquad (1.4)$$

M is the mass of the protein molecule in Da; N_o is Avogadro's number, 6.023×10^{23}; v_2 is the partial specific volume of the protein, typical value is 0.73 cm^3 g^{-1}; ρ is the density of solvent (1.0 g cm^{-3} for H_2O); η is the viscosity of the solvent (0.01 g cm^{-1} for H_2O).

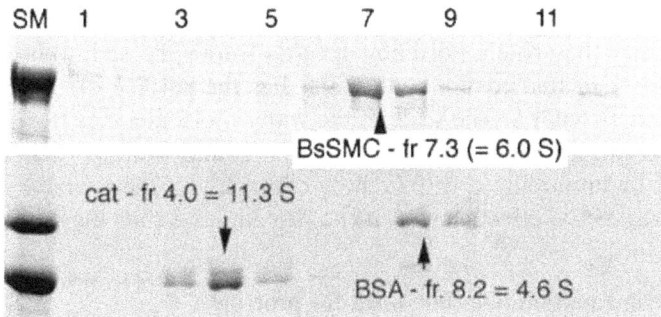

Figure 1.1. Glycerol gradient sedimentation analysis of SMC protein from *Bacillus subtilis* (BsSMC) (upper panel) and sedimentation standards catalase and bovine serum albumin (lower panel). A 200 μl sample was layered on a 5.0 ml gradient of 15%–40% glycerol in 0.2 M ammonium bicarbonate, and centrifuged in a Beckman SW55.1 swinging bucket rotor, 16 h, 38 000 rpm, 20 °C. 12 fractions of 400 μl each were collected from a hole in the bottom of the tube and each fraction was run on SDS–PAGE. Lane SM shows the starting material, and fraction 1 is the bottom of the gradient. The bottom panel shows that the 11.3 s catalase eluted precisely in fraction 4, while the 4.6 s BSA eluted mostly in fraction 8, with some in fraction 9. We estimated the BSA to be centered on fraction 8.2. Experiments with additional standard proteins have demonstrated that the 15%–40% glycerol gradients are linear over the range 3–20 s, so a linear interpolation is used to determine S of the unknown protein. BsSMC is in fractions 7 and 8, estimated more precisely at fraction 7.3. Extrapolating from the standards we determine a sedimentation coefficient of 6.0 s for BsSMC. Other experiments gave an average value of 6.3 s for BsSMC [19].

A critical factor in the equation is the frictional coefficient, f (dimensions g s^{-1}) which depends on both the size and shape of the protein. For a given mass of protein (or given volume), f will increase as the protein becomes elongated or asymmetrical. (f can be replaced by an equivalent expression containing R_s, the Stokes radius, to be discussed later.) S has the dimensions of time (seconds). For typical protein molecules S is in the range of 2–20×10^{-13} s, and the value 10^{-13} s is designated a Svedberg unit, S. Thus typical proteins have sedimentation coefficients of 2–20 s.

From the above definition of parameters it is clear that S depends on the solvent and temperature. In classical studies the solvent-dependent factors were eliminated and the sedimentation coefficient was extrapolated to the value it would have at 20 ° C in water (for which ρ and η are given above). This is referred to as $S_{20,w}$. In the present treatment we will be referring mostly to standard proteins that have already been characterized, or unknown ones that will be referenced to these in gradient sedimentation, so our use of S will always mean $S_{20,w}$.

A useful concept is the minimum value of f, which would obtain if the given mass of protein were packed into a smooth, unhydrated sphere. As we have discussed in section 1, the radius of this sphere will be $R_{min} = 0.066\ M^{1/3}$ (equation (1.2)). In about 1850 G G Stokes calculated theoretically the frictional coefficient of a smooth sphere (note that the equation is similar to that for the Stokes radius, to be discussed later, but the parameters here are different):

$$f_{min} = 6\pi\eta R_{min} \tag{1.5}$$

We have now designated f_{min} as the minimal frictional coefficient for a protein of a given mass, which would obtain if the protein were a smooth sphere of radius R_{min}.

The actual f of a protein will always be larger than f_{min} because of two things. First, the shape of the protein normally deviates from spherical, to be ellipsoidal or elongated; closely related to this is the fact that the surface of the protein is not smooth but rather rough on the scale of the water molecules it is traveling through. Second, all proteins are surrounded by a shell of bound water, 1–2 molecules thick, which is partially immobilized or frozen by contact with the protein. This water of hydration increases the effective size of the protein, and thus increases f.

1.4.1 The Perrin equation does not work for proteins

If one could determine the amount of water of hydration and factor this out, there would be hope that the remaining excess of f over f_{min} could be interpreted in terms of shape. Algorithms have been devised for estimating the amount of bound water from the amino acid sequence, but these generally do not distinguish between buried residues, which have no bound water, and surface residues which bind water. Some attempts have been made to base the estimate of bound water based on polar residues, which are mostly exposed on the surface. 0.3 g H_2O per g protein is a typical estimate, but in fact this kind of guess is almost useless for analyzing f.

In the older days, when there was some confidence in these estimates of bound water, physical chemists calculated a value called f_o, which was the frictional coefficient for a sphere that would contain the given protein, but enlarged by the estimated shell of water. (Other authors use f_o to designate what we term f_{min} [3, 4]. We recommend using f_{min} to avoid ambiguity.) The measured f for proteins was almost always larger than f_o, suggesting that the protein was asymmetrical or elongated. A very popular analysis was to model the protein as an ellipsoid of revolution, and calculate the axial ratio from f/f_o, using an equation first developed by Perrin. This approach is detailed in most classical texts of physical biochemistry. In fact the Perrin analysis always overestimates the asymmetry of the proteins, typically by a factor of two to five. It should not be used for proteins.

The problem is illustrated by an early collaborative study of phosphofructokinase, in which the laboratory of James Lee did hydrodynamics and our laboratory did EM [5]. We found by EM that the tetrameric particles were approximately cylinders, 9 nm in diameter and 14 nm long. The shape was therefore like a rugby ball, with an axial ratio of 1.5 for a prolate ellipsoid of revolution. The Lee group measured the molecular weight and sedimentation coefficient, determined f and estimated water of hydration and f_o. They then used the Perrin equation to calculate the axial ratio. The ratio was five, which would suggest that the protein had the shape of a hot dog. The EM structure (which was later confirmed by x-ray crystallography) shows that the Perrin equation overestimated the axial ratio by a factor of 3.

Teller *et al* [6] summarized the situation: 'Frequently the axial ratios resulting from such treatment are absurd in light of the present knowledge of protein structure.' They explained that the major problem with the Perrin equation is that it treats the protein as a smooth ellipsoid, when in fact the surface of the protein is quite rough. Teller *et al* went on to show how the frictional coefficient can actually be derived from the known atomic structure of the protein, by modeling the surface of the protein as a shell of small beads of radius 1.4 Å. The shell coated the surface of the protein, modeling its rugosity, and increasing the size of the protein by the equivalent of a single layer of bound water. This analysis has been extended by Garcia de la Torre and colleagues [7].

1.4.2 Interpreting shape from $f/f_{min} = S_{max}/S$

If the Perrin equation is useless, is there some other way that shape can be interpreted from f? The answer is yes, at a semiquantitative level. We have discovered simple guidelines where the ratio f/f_{min} can provide a good indication of whether a protein is globular, somewhat elongated or very elongated.

Instead of proceeding with the classical ratio f/f_{min}, where f is in non-intuitive units, we will reformulate the analysis directly in terms of the sedimentation coefficient, which is the parameter actually measured. We will define a value S_{max} as the maximum possible sedimentation coefficient, corresponding to f_{min}. S_{max} is the S value that would be obtained if the protein were a smooth sphere with no bound water. These two ratios are equal: $f/f_{min} = S_{max}/S$. Combining equations (1.2), (1.4) and (1.5), we have

$$S_{max} = 10^{13}M(1 - v_2\rho)/N_o(6\pi\eta R_{min}) = M[2.378 \times 10^{-4}]/R_{min} \qquad (1.6a)$$

Table 1.3. S_{max} calculated for proteins of different mass.

Protein M_r (kDa)	10	25	50	100	200	500	1000
S_{max} Svedbergs	1.68	3.1	4.9	7.8	12.3	22.7	36.1

Table 1.4. S_{max}/S values for representative globular and elongated proteins.

Globular Protein Standards Dimensions are from pdb files					
Protein	Dimensions nm	Mass	S_{max}	S	S_{max}/S
Phosphofructokinase	$14 \times 9 \times 9$	345 400	17.77	12.2	1.46
Catalase	$9.7 \times 9.2 \times 6.7$	230 000	13.6	11.3	1.20
Serum albumin	$7.5 \times 6.5 \times 4.0$	66 400	5.9	4.6	1.29
Hemoglobin	$6 \times 5 \times 5$	64 000	5.78	4.4	1.32
Ovalbumin	$7.0 \times 3.6 \times 3.0$	43 000	4.43	3.5	1.27
FtsZ	$4.8 \times 4 \times 3$	40 300	4.26	3.4	1.25
Elongated Protein Standards—Tenascin fragments [27, 28]; heat repeat [29, 30]					
Protein	Dimensions nm	Mass	S_{max}	S	S_{max}/S
TNfn1–5	$14.7 \times 1.7 \times 2.8$	50 400	4.94	3.0	1.65
TNfn1–8	$24.6 \times 1.7 \times 2.8$	78 900	6.64	3.6	1.85
TNfnALL	$47.9 \times 1.7 \times 2.8$	148 000	10.1	4.3	2.36
PR65/A HEAT repeat	$17.2 \times 3.5 \times 2.0$	60 000	5.53	3.6	1.54
fibrinogen	$46 \times 3 \times 6$	390 000	19.3	7.9	2.44

$$S_{max} = 0.00361 M^{2/3} \qquad (1.6b)$$

The leading factor of 10^{13} in (1.6a) converts S_{max} to Svedberg units. The numbers in brackets in (1.6a) are calculated using $v_2 = 0.73$ cm^3 g^{-1}, $\rho = 1.0$ g cm^{-3}, $\eta = 0.01$ g cm^{-1} s^{-1} $= 10^{-9}$ g nm^{-1} s^{-1}. The final expression, equation (1.6b) expresses S_{max} in Svedbergs for a protein of mass M in Daltons.

Some typical numerical values of S_{max} for proteins from 10 000 to 1 000 000 Da are given in table 1.3.

We have surveyed values of S_{max}/S for a variety of proteins of known structure. Table 1.4 presents S_{max}/S for a number of approximately globular proteins and for a range of elongated proteins, all of known dimensions. It turns out that S_{max}/S is an excellent predictor of the degree of asymmetry of a protein. From this survey of known proteins we can propose a number of general principals.

- No protein has $S_{max}/S = f/f_{min}$ smaller than ~1.2.
- For approximately globular proteins:
    S_{max}/S is typically between 1.2 and 1.3.
- For moderately elongated proteins:
    S_{max}/S is in the range of 1.5–1.9.

- For highly elongated proteins (tropomyosin, fibrinogen, extended fibronectin): S_{max}/S is in the range of 2.0–3.0.
- For very long, thread-like molecules like collagen, or huge extended molecules like the tenascin hexabrachion (not shown): S_{max}/S can range from 3 to 4 or more.

Apart from indicating the shape of a protein, S_{max}/S can often give valuable information about the oligomeric state, if one has some idea of the shape. For example, if one knows that the protein subunit is approximately globular (from EM for example), but finds $S_{max}/S = 2.1$, this would suggest that the protein in solution is a actually a dimer. On the other hand if one thinks a protein is a dimer, but finds $S_{max}/S < 1.0$ for the dimer mass, the protein is apparently sedimenting as a monomer.

The use of S_{max}/S to estimate protein shape has been described briefly in [8].

1.5 The Kirkwood/Bloomfield calculation

The understanding of how protein shape affects hydrodynamics is elegantly extended by an analysis originally developed by Kirkwood [9], and later extended by de la Torres and Bloomfield [10–12]. In its simplest application it calculates the sedimentation coefficient of a rigid oligomeric protein composed of subunits of known S and known spacing relative to each other. In more complex applications, a protein of any complex shape can be modeled as a set of non-overlapping spheres or beads. See Byron [13] for a comprehensive review of the principals and applications of hydrodynamic bead modeling of biological macromolecules.

The basis of the Kirkwood/Bloomfield analysis is to account for how each bead shields the others from the effect of solvent flow, and thereby determine the hydrodynamics of the ensemble from its component beads. Figure 1.2 shows a simple example of the bead modeling approach, and provides an instructive look at how size and shape affect sedimentation. There are several important conclusions.

- A rod of three beads has about a two-fold higher S than a single bead.
- S_{max}/S is 1.18 for the single bead (the effect of the assumed shell of water); 1.34 for the three-bead rod; 1.93 for the straight 11-bead rod. This is consistent with the principals given in section 3 for globular, somewhat elongated and very elongated particles.
- Bending the rod at 90° in the middle causes only a small increase in S. Bending it into a U-shape with the arms about one bead diameter apart increases S a bit more. Bending this same 11-bead structure more sharply, so the two arms are in contact, causes a substantial increase in S, from 5.05 to 5.58. The guiding principle is that folding affects S when one part of the molecule is brought close enough to another to shield it from water flow.

1.6 Gel filtration chromatography and the Stokes radius

'Gel filtration chromatography is widely used for determining protein molecular weight.' This quote from Sigma–Aldrich bulletin 891A is a widely held

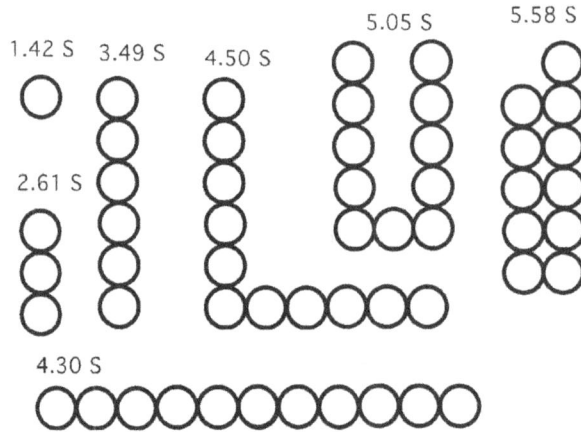

Figure 1.2. Each bead models a 10 kDa domain, with an assumed sedimentation coefficient of 1.42 s. The radius of the bead is 1.67 nm, using $R_{min} = 1.42$ nm, and adding 0.25 nm for a shell of water. The beads are an approximation to FN-III or Ig domains, which are ~1.7 × 2.8 × 3.5 nm. The sedimentation coefficients of multi-bead structures were calculated by the formula of Kirkwood/Bloomfield.

misconception. The fallacy is obscurely corrected by a later note in the bulletin that 'Once a calibration curve is prepared, the elution volume for a protein of similar shape, but unknown weight, can be used to determine the MW.' The key issue is 'of similar shape.' Generally the calibration proteins are all globular, and if the unknown protein is also globular the calibrated gel filtration column does give a good approximation of its molecular weight. The problem is that the shape of an unknown protein is generally unknown. If the unknown protein is elongated it can easily elute at a position twice the molecular weight of a globular protein.

The gel filtration column actually separates proteins not on their molecular weight, but on their frictional coefficient. Since the frictional coefficient, f, is not an intuitive parameter, it is usually replaced by the Stokes radius R_s. R_s is defined as the radius of a smooth sphere that would have the actual f of the protein. This is much more intuitive since it allows one to imagine a real sphere approximately the size of the protein, or somewhat larger if the protein is elongated and has bound water.

As mentioned above for equation (1.5), Stokes calculated theoretically the frictional coefficient of a smooth sphere to be:

$$f = 6\pi\eta R_s \tag{1.7}$$

The Stokes radius R_s is larger than R_{min} because it is the radius of a smooth sphere whose f would match the actual f of the protein. It accounts for both the asymmetry of the protein and the shell of bound water. More quantitatively, $f/f_{min} = S_{max}/S = R_s/R_{min}$.

Siegel and Monte [4] argued convincingly that the elution of proteins from a gel filtration column correlates closely with the Stokes radius, R_s, presenting experimental data from a wide range of globular and elongated proteins. The Stokes radius is known for large number of proteins, including ones convenient for

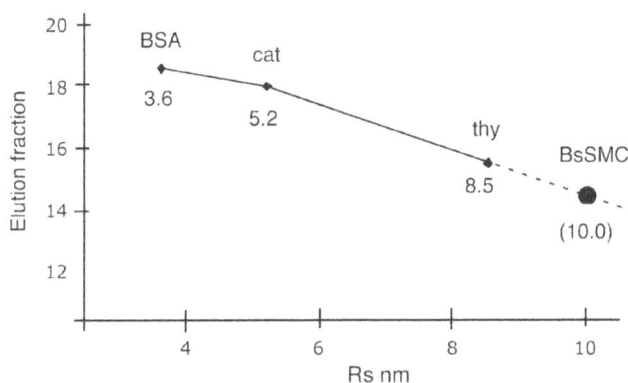

Figure 1.3. Determination of R_s of BsSMC by gel filtration. The column was calibrated by running standard proteins BSA, catalase and thyroglobulin over the column, then BsSMC. BsSMC eluted in fraction 14.2, which corresponds to an R_s of 10 nm on the extrapolated curve. In repeated experiments the average R_s was 10.3 nm [19].

calibrating gel filtration columns (table 1.5). Figure 1.3 shows an example where the R_s of the unknown protein BsSMC was determined by gel filtration.

The standard proteins should span R_s values above and below that of the protein of interest (but in the case of BsSMC a short extrapolation to a larger value was used). The literature generally recommends determining the void and included volumes of the column and plotting a partition coefficient K_{AV} [4]. However, we have found it generally satisfactory to simply plot elution position versus R_s for the standard proteins. This generally gives an approximately linear plot, but otherwise it is satisfactory to draw lines between the points and read the R_s of the protein of interest from its elution position on this standard curve.

A gel filtration column can determine R_s relative to the R_s of the standard calibration proteins. The R_s of these standards was generally determined from experimentally measured diffusion coefficients. Some tabulations of hydrodynamic data list the diffusion coefficient, D, rather than R_s, so it is worth knowing the relationship:

$$D = kT/f = kT / (6\pi\eta R_s) \qquad (1.8)$$

where $k = 1.38 \times 10^{-16}$ g cm^2 s^{-2} K^{-1} is Boltzman's constant and T is the absolute temperature. k is given here in cgs units because D is typically expressed in cgs; R_s will be expressed in cm in this equation. Typical proteins have D in the range of 10^{-6}–10^{-7} cm^2 s^{-1}. Converting to nm and for $T = 300$ K and $\eta = 0.01$:

$$R_s = (1/D) 2.2 \times 10^{-6}, \qquad (1.9)$$

where R_s is in nm and D is in cm^2 s^{-1}.

Simply knowing R_s is not very valuable in itself, except for estimating the degree of asymmetry, but this would be the same analysis developed above for S_{max}/S. However, if one determines both R_s and S, this permits a direct determination of molecular weight, which cannot be deduced from either one alone. This is described in the next section.

1.7 Determining the molecular weight of a protein molecule—combining S and R_s à la Siegel and Monte

With the completion of multiple genomes and increasingly good annotation, the primary sequence of almost any protein can be found in the databases. The molecular weight of every protein subunit is therefore known from its sequence. But an experimental measure is still needed to determine if the native protein in solution is a monomer, dimer or oligomer, or if it forms a complex with other proteins. If one has a purified protein the molecular weight can be determined quite accurately by sedimentation equilibrium in the analytical ultracentrifuge. This technique has made a strong comeback with the introduction of the Beckman XL-A analytical ultracentrifuge. There are a number of good reviews [14, 15], and the documentation and programs that come with the centrifuge are very instructive.

What if one does not have an XL-A centrifuge, or the protein of interest is not purified? In 1966, Siegel and Monte [4] proposed a method that achieves the results of sedimentation equilibrium, with two enormous advantages. First, it requires only a preparative ultracentrifuge for sucrose or glycerol gradient sedimentation, and a gel filtration column. This equipment is available in most biochemistry laboratories. Second, the protein of interest need not be purified; one needs only an activity or an

Table 1.5. Standards for hydrodynamic analysis.

Protein	M_r aa seq	$S_{20,w}$	Smax/S	R_s(nm)	Source	M_r S–M
Ribonuclease A beef pancreas	14 044	2.0[a]	1.05[a]	1.64	HBC	13 791
Chymotrypsinogen A beef pancreas	25 665	2.6	1.21	2.09	HBC	22 849
Ovalbumin hen egg	42 910 s	3.5	1.27	3.05	HBC	44 888
Albumin beef serum	69 322	4.6[a]	1.33	3.55	S–M,HBC	68 667
Aldolase rabbit muscle	157 368	7.3	1.45	4.81	HBC	147 650
Catalase beef liver	239 656	11.3	1.21	5.2	S–M	247 085
Apo-ferritin horse spleen	489 324	17.6	1.28	6.1	HBC	451 449
Thyroglobulin bovine	606 444	19	1.37	8.5	HBC	679 107
Fibrinogen, human	387 344	7.9	2.44	10.7	S–M	355 449

Gel filtration calibration kits, containing globular proteins of known molecular weight and R_s, are commercially available (GE healthcare, Sigma–Aldrich). These same proteins can be used for sedimentation standards. The proteins in these kits are included in the table along with some others that we have found useful. The values for M_r given in the first column are from amino acid sequence data. Values for $S_{20,w}$ and R_s are from the Siegel–Monte paper (indicated S–M under source), or the *CRC Handbook of Biochemistry* [3] (indicated HBC). They agree with the values listed for R_s in the GE Healthcare gel filtration calibration kit, with the exception of ferritin. The 'M_r *calc*' in the last column was obtained by our simplification of the Siegel–Monte calculation ($M = 4205\ s\ R_s$). Note that the worst disagreement with 'M_r aa seq' is about 10%.
[a] S for ribonuclease A is questionable because of the low S_{max}/S (1.05). S values for bovine serum albumin vary in the literature from 4.3 to 4.9. Many sources use 4.3, but we find that 4.6 gives a better fit with other standards (note that the standard curve in figure 1.5 used 4.3, but 4.6 would have placed it closer to the line).

antibody to locate it in the fractions. This is a very powerful technique, and should be in the repertoire of every protein biochemist.

The methodology is very simple. The protein is run over a calibrated gel filtration column to determine R_s, and hence f. Separately the protein is centrifuged through a glycerol or sucrose gradient to determine S. One then uses the Svedberg equation (equation (1.4)) to obtain M as a function of R_s and S.

$$M = SN_o(6\pi\eta R_s)/ (1 - v_2\rho) \tag{1.10a}$$

setting $\eta = 0.01$, $v_2\rho = 0.73$, converting S to Svedberg units and R_s to nm, we can simplify further:

$$M = 4205 \, (SR_s) \tag{1.10b}$$

where S is in Svedberg units, R_s is in nm and M is in Daltons.

This is pretty simple! Importantly, in typical applications this method gives the protein mass within about $\pm 10\%$. This is more than enough precision to distinguish between monomer, dimer or trimer.

Application to BsSMC. In the sections above we showed how S of the SMC protein from *B. subtilis* was determined to be 6.3 Svedberg units from glycerol gradient sedimentation, and R_s was 10.3 nm, from gel filtration. Putting these values in equation (1.10b) we find that the molecular weight of BsSMC is 273 000 Da. From the amino acid sequence we know that the molecular weight of one BsSMC subunit is 135 000 Da. The Siegel–Monte analysis finds that the BsSMC molecule is a dimer.

Knowing that BsSMC is a dimer with molecular weight 270 000 Da, we can now determine its S_{max}/S. S_{max} is 15.1 (equation (1.6b)) so S_{max}/S is 2.4. The BsSMC molecule is thus expected to be highly elongated. EM (see below) confirmed this prediction.

1.8 Electron microscopy of protein molecules

Since the early 1980s electron microscopy has become a powerful technique for determining the size and shape of single protein molecules, especially ones larger than 100 kDa. Two techniques available in most EM laboratories, rotary shadowing and negative stain, can be used for imaging single molecules. Cryo EM is becoming a powerful tool for protein structural analysis, but it requires special equipment and expertise. For a large number of applications rotary shadowing and negative stain provide the essential structural information.

For rotary shadowing a dilute solution of protein is sprayed on mica, the liquid is evaporated in a high vacuum, and platinum metal is evaporated onto the mica at a shallow angle. The mica is rotated during this process, so the platinum builds up on all sides of the protein molecules. The first EM images of single protein molecules were obtained by Hall and Slayter using rotary shadowing [16]. Their images of fibrinogen showed a distinctive trinodular rod. However, rotary shadowing fell into disfavor because the images were difficult to reproduce. Protein tended to aggregate and collect salt, rather than spread as single molecules. In 1976 James Pullman, then a graduate student at the University of Chicago, devised a protocol with one simple but crucial modification—he added 30% glycerol to the protein solution. For reasons that are still not understood, the glycerol greatly helps the spreading of the protein as single molecules.

Pullman never published his protocol, but two labs saw his mimeographed notes and tested out the effect of glycerol, as a part of their own attempts to improve rotary shadowing [17, 18]. They obtained reproducible and compelling images of fibrinogen (the first since the original Hall and Slayter study, and confirming the trinodular rod structure) and spectrin (the first ever images of this large protein). The technique has since been used in characterizing hundreds of protein molecules.

Figure 1.4 shows rotary shadowed BsSMC, fibrinogen and hexabrachion (tenascin). BsSMC is highly elongated, consistent with its high S_{max}/S discussed above [19]. The fibrinogen molecules show the trinodular rod, but these images also resolved a small fourth nodule next to the central nodule [20], not seen in earlier studies. The central nodule is about 50 kDa, and the smaller fourth nodule is about 20 kDa. The 'hexabrachion' tenascin molecule [21] illustrates the power of rotary shadowing at two extremes. First, the molecule is huge. Each of its six arms is made up of ~30 repeating small domains, totaling ~200 000 Da. At the larger scale the EM shows that each arm is an extended structure, matching the length expected if the repeating domains are an extended string of beads. At the finer scale, the EM can distinguish the different sized domains. The inner segment of each arm is a string of 3.5 kDa EGF domains, seen here as a thinner segment. A string of 10 kDa FN-III domains is clearly distinguished as a thicker outer segment. The terminal knob is a single, 22 kDa fibrinogen domain. The R_{min} of these domains are 0.8, 1.7 and 2.8 nm, and these can be distinguished by rotary shadowing. Rotary shadowing EM can visualize single globular domains as small as 10 kDa (3.5 nm diameter), and elongated molecules as thin as 1.5 nm (collagen).

Negative stain is another EM technique capable of imaging single protein molecules. It is especially useful for imaging larger molecules with a complex internal structure, which appear only as a large blob in rotary shadowing. Importantly, non-covalent protein–protein bonds are sometimes disrupted in the rotary shadowing technique [8], but uranyl acetate, in addition to providing high

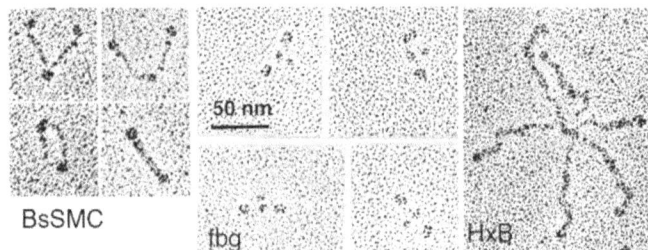

BsSMC   50 nm   fbg   HxB

Figure 1.4. Rotary shadowing EM of three highly elongated protein molecules: the SMC protein from *B. subtilis* [19], fibrinogen [20], and the hexabrachion protein, tenascin [21]. Reprinted with permission from the indicated references.

resolution contrast, fixes oligomeric protein structures in a few milliseconds [22]. An excellent review of modern techniques of negative staining, with comparison to cryo EM, is given in [23].

The simple picture of the molecule produced by EM is frequently the most straightforward and satisfying structural analysis at the 1–2 nm resolution. When the structure is confirmed by hydrodynamic analysis the interpretation is even more compelling.

1.9 Hydrodynamic analysis and EM applied to large multi-subunit complexes

The text box above showed the application of the Siegel–Monte analysis to BsSMC, which had only one type subunit and was found to be a dimer. Similar hydrodynamic analysis can be used to analyze multi-subunit protein complexes. There are many examples in the literature; I will show here an elegant application to DASH/Dam1.

The protein complex called DASH or Dam1 is involved in attaching chromosomal kinetochores to microtubules in yeast. DASH/Dam1 is a complex of ten proteins that assemble into a particle containing one copy of each subunit. These complexes further assemble into rings that can form a sliding washer on the microtubule [24, 25]. The basic ten-subunit complex has been purified from yeast, and has also been expressed in *E. coli* and purified (this required the heroic effort of expressing all ten proteins simultaneously [24]). Figure 1.5 shows the hydrodynamic characterization of the purified protein complex, and illustrates several important features.

- For both the gel filtration (size exclusion chromatography, figure 1.5(a)) and gradient sedimentation, figure 1.5(b), two calibration curves of known protein standards are shown, green and black. These are independent calibration runs. In this study the gel filtration column was calibrated in terms of the reciprocal diffusion coefficient, $1/D$, which is proportional to R_s (equation (1.7)).
- The fractions were analyzed by western blot for the location of two proteins of the complex, Spc34p and Hsk3p. Methods notes that 1 ml fractions from gel filtration were precipitated with perchloric acid and rinsed with acetone prior to SDS–PAGE, an essential amplification for the dilute samples of yeast cytoplasmic extract. These two proteins eluted together in both gel filtration and sedimentation, consistent with their being part of the same complex.
- The profiles of the two proteins were identical when analyzed in their native form in yeast cytoplasmic extract, and as the purified complex expressed in *E. coli*. This is strong evidence that the expression protein is correctly folded and assembled.
- There is minimal trailing of any subunits. This means that there is no significant dissociation during the tens of minutes for the gel filtration, or the 12 h centrifugation. The complex is held together by very high affinity bonds, making it essentially irreversible.
- Combining the $R_s = 7.6$ nm from $1/D = 0.35 \times 10^{-7}$, and $S = 7.4$, equation (1.10b) gives a mass of $M = 236$ kDa, close to the 204 kDa obtained from adding the mass of the ten subunits. S_{max} is 12.6 giving $S_{max}/S = 1.7$, suggesting a moderately elongated protein.

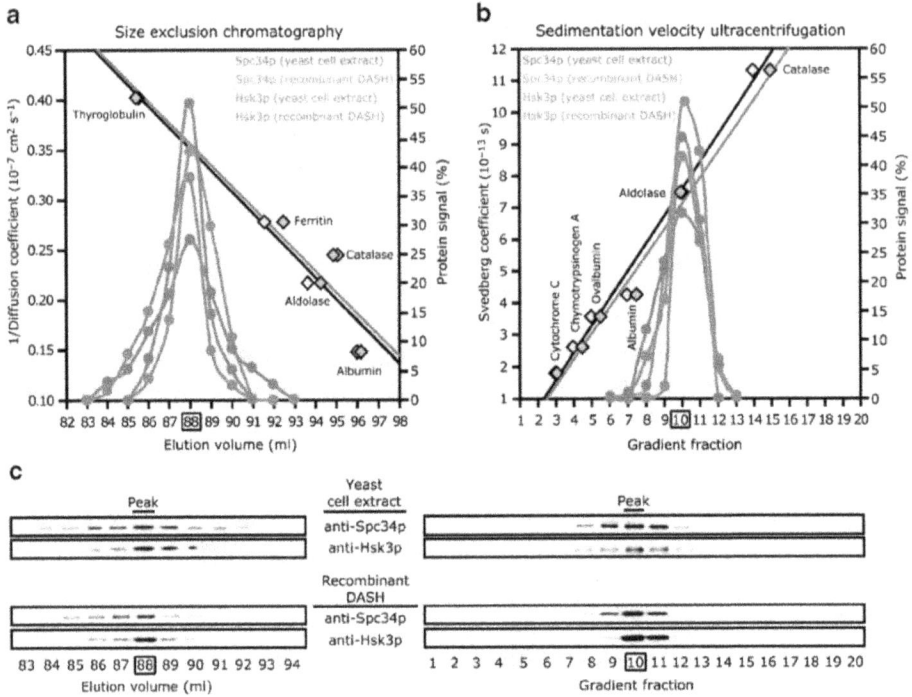

Figure 1.5. Hydrodynamic analysis of the DASH/Dam1 complex. Gel filtration is shown in a and sucrose gradient sedimentaion in b. Independent calibration curves using standard proteins are shown in black and green. Dark and light blue show Spc34p in yeast cytoplasmic extract and in the purified recombinant protein. Red and purple show Hsk3p. The proteins were identified and quantitated by western blot of the fractions, shown in c. The four protein bands eluted together at $1/D = 0.35 \times 10^7$, corresponding to $R_s = 7.6$ nm, and at 7.4 S. Reproduced from Miranda *et al* [24] with permission from Springer Nature.

Figure 1.6 shows EM images of DASH/DAM1 by rotary shadowing (figure 1.6a) and negative stain (figure 1.6b). Rotary shadowing showed irregular particles about 13 nm in diameter [24]. The particles had variable and frequently elongated shapes, but internal structure could not be resolved. A later study used state of the art negative staining and sophisticated computer programs to sort images into classes and average them [26]. These images resolved a complex internal structure. The negative stain showed most of the particles (80%) to be dimers, with 15% monomers and 5% trimers. This contradicts the hydrodynamic analysis of Miranda *et al* [24] showing that the particles were monomers. The reason for this discrepancy is not known.

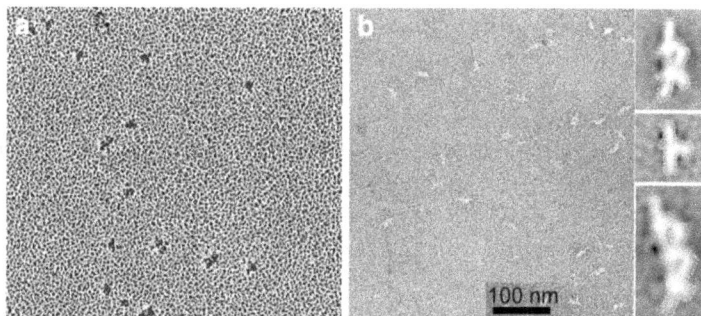

Figure 1.6. EM of DASH/Dam1. (a) Rotary shadowing shows particles roughly 13 nm in size, with irregular shape. (b) State of the art negative stain coupled with single particle averaging shows a complex internal structure of the elongated particles. The scale bar indicates 100 nm for the unprocessed images. The averaged images on the right show a monomer, dimer and trimer. These panels are 14 nm wide. The dimer was the predominant species. Left panel (rotaty shadowing) from Miranda *et al* [24] reprinted with permission from Springer Nature. Right panels (negative stain) reprinted with permission of Wang *et al* [26].

Acknowledgement: Supported by NIH grant CA47056.

References

[1] Richards F M 1974 The interpretation of protein structures: total volume, group volume distributions and packing density *J. Mol. Biol.* **82** 1–14

[2] Gittes F, Mickey B, Nettleton J and Howard J 1993 Flexural rigidity of microtubules and actin filaments measured from thermal fluctuations in shape *J. Cell Biol.* **120** 923–34

[3] Sober H A 1966 *Handbook of Biochemistry* (Cleveland, OH: The Chemical Rubber Co.)

[4] Siegel L M and Monte K J 1966 Determination of molecular weights and frictional ratios of proteins in impure systems by use of gel filtration and density gradient centrifugation *Biochim. Biophys. Acta* **112** 346–62

[5] Hesterberg L K, Lee J C and Erickson H P 1981 Structural properties of an active form of rabbit muscle phosphofructokinase *J. Biol. Chem.* **256** 9724–30

[6] Teller D C, Swanson E and De Haen C 1979 The translational friction coefficient of proteins *Methods Enzymol.* **61** 103–24

[7] Garcia De La Torre J, Huertas M L and Carrasco B 2000 Calculation of hydrodynamic properties of globular proteins from their atomic-level structure *Biophys. J.* **78** 719–30

[8] Schürmann G, Haspel J, Grumet M and Erickson H P 2001 Cell adhesion molecule L1 in folded (horseshoe) and extended conformations *Mol. Biol. Cell* **12** 1765–73

[9] Kirkwood J G 1954 The general theory of irreversible processes in solutions of macro-molecules *J. Polymer Sci.* **12** 1–14

[10] Bloomfield V, Dalton W O and van Holde K E 1967 Frictional coefficients of multisubunit structures. I. Theory *Biopolymers* **5** 135–48

[11] Carrasco B and Garcia de la Torre J 1999 Hydrodynamic properties of rigid particles: comparison of different modeling and computational procedures *Biophys. J.* **76** 3044–57

[12] Garcia de la Torre J, Llorca O, Carrascosa J L and Valpuesta J M 2001 HYDROMIC: prediction of hydrodynamic properties of rigid macromolecular structures obtained from electron microscopy images *Eur. Biophys. J.* **30** 457–62

[13] Byron O 2008 Hydrodynamic modeling: the solution conformation of macromolecules and their complexes *Methods Cell. Biol.* **84** 327–73

[14] Schuster T M and Toedt J M 1996 New revolutions in the evolution of analytical ultracentrifugation *Curr. Opin. Struct. Biol.* **6** 650–58

[15] Hansen J C, Lebowitz J and Demeler B 1994 Analytical ultracentrifugation of complex macromolecular systems [Review] *Biochemistry* **33** 13155–63

[16] Hall C E and Slayter H S 1959 The fibrinogen molecule: its size, shape, and mode of polymerization *J. Biophys. Biochem. Cytol.* **5** 11–6

[17] Fowler W E and Erickson H P 1979 Trinodular structure of fibrinogen. Confirmation by both shadowing and negative stain electron microscopy *J. Mol. Biol.* **134** 241–49

[18] Shotton D M, Burke B E and Branton D 1979 The molecular structure of human erythrocyte spectrin *J. Mol. Biol.* **131** 303–29

[19] Melby T E, Ciampaglio C N, Briscoe G and Erickson H P 1998 The symmetrical structure of structural maintenance of chromosomes (SMC) and MukB proteins: long, antiparallel coiled coils, folded at a flexible hinge *J. Cell Biol.* **142** 1595–604

[20] Erickson H P and Fowler W E 1983 Electron microscopy of fibrinogen, its plasmic fragments and small polymers *Ann. N.Y. Acad. Sci.* **408** 146–63

[21] Erickson H P and Iglesias J L 1984 A six-armed oligomer isolated from cell surface fibronectin preparations *Nature* **311** 267–69

[22] Zhao F Q and Craig R 2003 Capturing time-resolved changes in molecular structure by negative staining *J. Struct. Biol.* **141** 43–52

[23] Ohi M, Li Y, Cheng Y and Walz T 2004 Negative staining and image classification—powerful tools in modern electron microscopy *Biol. Proced. Online* **6** 23–34

[24] Miranda J J, De Wulf P, Sorger P K and Harrison S C 2005 The yeast DASH complex forms closed rings on microtubules *Nat. Struct. Mol. Biol.* **12** 138–43

[25] Westermann S, Avila-Sakar A, Wang H W, Niederstrasser H, Wong J, Drubin D G, Nogales E and Barnes G 2005 Formation of a dynamic kinetochore–microtubule interface through assembly of the Dam1 ring complex *Mol. Cell* **17** 277–90

[26] Wang H W, Ramey V H, Westermann S, Leschziner A E, Welburn J P, Nakajima Y, Drubin D G, Barnes G and Nogales E 2007 Architecture of the Dam1 kinetochore ring complex and implications for microtubule-driven assembly and force-coupling mechanisms *Nat. Struct. Mol. Biol.* **14** 721–26

[27] Aukhil I, Joshi P, Yan Y and Erickson H P 1993 Cell- and heparin-binding domains of the hexabrachion arm identified by tenascin expression proteins *J. Biol. Chem.* **268** 2542–53

[28] Leahy D J, Axel R and Hendrickson W A 1992 Crystal structure of a soluble form of the human T cell coreceptor CD8 at 2.6 Å resolution *Cell* **68** 1145–62

[29] Chen S C, Kramer G and Hardesty B 1989 Isolation and partial characterization of an M_r 60,000 subunit of a type 2A phosphatase from rabbit reticulocytes *J. Biol. Chem.* **264** 7267–75

[30] Groves M R, Hanlon N, Turowski P, Hemmings B A and Barford D 1999 The structure of the protein phosphatase 2A PR65/A subunit reveals the conformation of its 15 tandemly repeated HEAT motifs *Cell* **96** 99–110

IOP Publishing

Principles of Protein–Protein Association

Harold P Erickson

Chapter 2

Basic thermodynamics of reversible association

Although many enzymes function as a single protein subunit, the majority of protein molecules associate with other protein molecules to achieve their functions. Protein subunits are held together by interfaces that we will refer to as protein–protein bonds. Some interfaces are stabilized by covalent bonds, namely disulfide bonds or the more specialized transglutaminase or lysyl oxidase bonds. These covalent bonds occur primarily or exclusively in extracellular proteins. Cytoplasmic protein–protein bonds are non-covalent and reversible, as are extracellular binding of growth factors to their receptors. We will focus on this reversible association as the primary protein–protein bond.

The reversible association of proteins can be characterized by an equilibrium constant or the associated free energy. Consider the association of two subunits to form a dimer. We will use G to describe one subunit (think of it as a growth factor or ligand) and R for the other (its receptor, which is typically present at far lower concentration than G). We will ignore kinetics for now and describe the reaction after it has reached equilibrium.

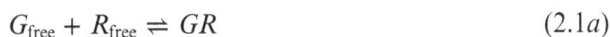

$$G_{\text{free}} + R_{\text{free}} \rightleftharpoons GR \qquad (2.1a)$$

$$K_{\text{D}} = \frac{[G_{\text{free}}][R_{\text{free}}]}{[GR]} = 1/K_{\text{A}} \qquad (2.1b)$$

The equilibrium is defined by a constant K (capital K is used for equilibrium constants, small k for kinetics). The square brackets indicate the concentration in units of molar, e.g. $[GR]$ is the concentration of receptor-growth factor complex, and $[R_{\text{free}}]$ is the concentration of unbound receptor. There are two ways to express K. K_{D} is the dissociation equilibrium constant, and its units are M. This is convenient because it expresses the strength of a reaction in units that can be compared directly with the concentration of the reactants. A complication of K_{D} is that it is typically a small number (e.g. $K_D = 10^{-6}$ M), and it becomes an even smaller number for a stronger bond (e.g. $K_D = 10^{-8}$ M is a stronger bond than $K_D = 10^{-6}$ M). Its inverse K_A, the association equilibrium constant, is more intuitive for discussing bond strengths because a stronger bond means a bigger K_A. However, K_A has the less intuitive unit of M^{-1}.

An important ratio is the concentration of liganded receptors $[GR]$, over the concentration of unliganded receptors $[R_{free}]$. We can rearrange the above equation:

$$\frac{[GR]}{[R_{free}]} = \frac{[G_{free}]}{K_D} \tag{2.2}$$

This relationship is exact, but it is complicated by the fact that $[G_{free}]$ and $[R_{free}]$ are the concentrations of G and R that remain free at equilibrium, which vary with the reaction. What we really want is a relationship in terms of $[G_{TOT}]$ and $[R_{TOT}]$, the total amount of G and R that the experimenter added to the reaction. To achieve this we can substitute $[G_{free}] = [G_{TOT}] - [GR]$ and $[R_{free}] = [R_{TOT}] - [GR]$ and solve for the concentration of GR.

$$[GR] = \frac{[G_{free}][R_{free}]}{K_D} = \frac{([R_{TOT}] - [GR])([G_{TOT}] - [GR])}{K_D} \tag{2.3}$$

$$K_D[GR] = [R_{TOT}][G_{TOT}] - [GR]([R_{TOT}] + [G_{TOT}]) + [GR]^2 \tag{2.4}$$

$$0 = [GR]^2 - [GR]([R_{TOT}] + [G_{TOT}] + K_D) + [R_{TOT}][G_{TOT}] \tag{2.5}$$

This is a quadratic equation whose solution is:

$$[GR] = \frac{([R_{TOT}] + [G_{TOT}] + K_D) - \sqrt{([R_{TOT}] + [G_{TOT}] + K_D)^2 - 4[R_{TOT}][G_{TOT}]}}{2} \tag{2.6}$$

This is a rather daunting formula. For a typical experiment you have a single fixed concentration $[R_{TOT}]$ and data points measuring $[GR]$ over a broad range of $[G_{TOT}]$. K_D is the unknown parameter, and you want to find the value of K_D that gives the best fit to the data. It turns out this is a simple task with modern computers and an algorithm for 'nonlinear fit.' Probably most accessible is the 'solver' routine in Excel (see appendix for a brief tutorial on how to do this). Mathematica or Maple are good alternatives. These routines will find the K_D that gives the best fit to your experimental data.

A simplified interpretation of K_D

The analysis is greatly simplified if G is present in great excess over R. This will generally be true for growth factors and receptors. In this case the amount of G that is lost in the complex is negligible, and $[G_{free}] \approx [G_{TOT}]$. Then we have a simple relation for the fraction of R in complex:

$$\frac{[GR]}{[R_{\text{free}}]} \approx \frac{[G_{\text{TOT}}]}{K_D} \quad \text{(for } [R_{\text{TOT}}] \ll [G_{\text{TOT}}]) \tag{2.7}$$

From this relation we see that the fraction of receptor occupied is determined simply by the ratio of $[G_{\text{TOT}}]$ to K_D.

Now we have three regimes:

If $[G_{\text{TOT}}] \ll K_D$, then GR complexes will be rare; most of the R will remain unliganded.
If $[G_{\text{TOT}}] = K_D$, then R will be 50% empty and 50% liganded.
If $[G_{\text{TOT}}] \gg K_D$, then most of the R will be liganded.

And we can specify bonds as high, moderate or weak affinity according to K_D:

K_D = nM to pM (10^{-9} to 10^{-12} M) are high affinity (receptor-growth factor)
K_D = µM (10^{-6} M) are moderate affinity (actin or tubulin onto polymer; Ab binding antigen)
K_D = mM (10^{-3} M) are weak affinity (Hb in rbc is 0.005 M and does not self-associate. Therefore K_D for Hb–Hb is $\gg$ 0.005 M).

The dissociation equilibrium constant, K_D, is especially useful because it allows a quick comparison with the molar concentration of the reactants. However, the association constant, $K_A = 1/K_D$, has the intuitive advantage that a stronger association corresponds to a larger K_A. For most of this chapter, and in the chapter on cooperativity, we will use the association constant, K_A.

Free energy, entropy, and intrinsic bond energy
The equilibrium constant is related to the free energy by these standard equations:

$$\Delta G_A = -RT \ln(K_A); \tag{2.8a}$$

$$K_A = e^{-\Delta G_A/RT} \tag{2.8b}$$

Note that favorable association, a large K_A, means a negative free energy of association. This may sound counterintuitive, but it is the way thermodynamics was developed. (The reader may also be bothered by taking the ln of K_A, which has units M^{-1}. The ΔG_A is actually referenced to a standard state, which is conveniently taken to be 1 M. So consider the K_A in equation (2.3) to be divided by 1 M^{-1}, which gets rid of the units.)

This free energy describes all of the chemical and energetic factors involved in the association reaction. It is useful to break this term down into two opposing free energies, one favoring association and one opposing it. The importance of explicitly separating these two terms has been recognized since the work of Doty and Myers on insulin dimerization in 1953 [1]; it has been rediscovered about every ten years since; it was a crucial point in the analysis of Chothia and Janin, 1975 [2]; and it provides the basis for the analysis of cooperativity (chapter 7 and [3]).

The two free energy terms are the *intrinsic bond energy*, which includes all the chemical forces acting across the subunit interface, and the *intrinsic subunit entropy*, an entropic factor that opposes association.

$$\Delta G_A = \Delta G_{bond} + \Delta G_S; \qquad (2.9a)$$

$$\Delta G_{bond} = \Delta G_A - \Delta G_S \qquad (2.9b)$$

ΔG_A is the net free energy related to the association constant in equation (2.8a, b). ΔG_{bond} is the intrinsic bond energy. It includes all the chemical forces intrinsic to the protein–protein bond. ΔG_{bond} is a negative number because it favors the association; the stronger the bond, the larger its absolute value. The Chothia and Janin analysis discussed in the next chapter is directed at estimating this intrinsic bond energy and determining all of the chemical forces that contribute to it.

The term ΔG_s is the intrinsic subunit entropy (S), expressed here in units of free energy ($T\Delta S$). Think of this as the free energy required to immobilize a subunit in a dimer or polymer, independent of the strength or number of bonds formed. Free energy is required because entropy is lost when the subunit is immobilized. Before dimer formation each subunit has three degrees of translational and three degrees of rotational freedom. In the dimer one subunit still retains its six degrees of freedom, but the other subunit has lost its independent translation and rotation. The question is, how much free energy does it take to immobilize a subunit, to compensate for the loss of three translational and three rational degrees of freedom? The best current value is $\Delta G_s = +6$ kcal mol^{-1}.

Chothia and Janin estimated ΔG_s to be 20–30 kcal mol^{-1}, based on a quantum mechanical calculation [2]. Erickson noted that proteins in a dimer would retain motions corresponding to ± 1 Å, and calculated a $\Delta G_s = 11$ kcal mol^{-1} [3]. However, he noted that this value was too high to fit a reasonable model of actin nucleation, and suggested 7 kcal mol^{-1} as a reasonable upper limit. Later, Horton and Lewis plotted ΔG values for a range of protein dimers, and the intercept gave a value of $\Delta G_s = 6$ kcal mol^{-1} [4], which we will use here. An important feature of ΔG_s is that it depends very little on the size or shape of the protein subunit. That is because these features enter the calculation as the logarithm. Therefore, the $\Delta G_s = 6$ kcal mol^{-1} can be considered universal for immobilizing a protein subunit in a dimer, oligomer or polymer.

It is useful to think about the intrinsic free energy as an entropy tax. The tax is a flat rate, not progressive. The entropy tax must be paid once for any association, and the tax is the same regardless of the size of the subunit and the strength of the bond. Another way to think about equations (2.9a, b) is that the intrinsic bond energy must be sufficient to achieve the observed K_A, and it must also pay the entropy tax of 6 kcal mol^{-1}.

$$\Delta G_{bond} = \Delta G_A - 6 \, \text{kcal mol}^{-1} \tag{2.10}$$

Let's put this in perspective and illustrate the calculations with some numbers. A typical modest protein–protein association, such as actin assembly or a weak antibody, will have a $K_D = 10^{-6}$ M ($K_A = 10^6$ M^{-1}). In equations (2.2)–(2.8), R has the value 2 cal deg^{-1}.mol, and T is the absolute temperature = 300 degrees Kelvin (at 27 °C, chosen here to make RT a round 600 cal mol^{-1} = 0.6 kcal mol^{-1}). Consider the case of actin, where $K_A \sim 10^6$ M^{-1}.

$$\Delta G_A = -RT \ln(K_A) = -0.6 \ln(10^6) = -8.3 \, \text{kcal mol}^{-1} \tag{2.11}$$

$$\Delta G_{bond} = \Delta G_A - 6 \, \text{kcal mol}^{-1} = -8.3 - 6 = -14.3 \, \text{kcal mol}^{-1}. \tag{2.12}$$

The intrinsic bond energy ΔG_{bond} must be larger than ΔG_A because it must compensate for the entropic energy loss.

Now let's consider a higher affinity interaction, $K_A = 10^9$ M^{-1} (this would be typical for a growth factor binding its receptor).

$$\Delta G_A = -RT \ln(K_A) = -0.6 \ln(10^9) = -12.4 \, \text{kcal mol}^{-1} \tag{2.13}$$

$$\Delta G_{bond} = \Delta G_A - 6 \, \text{kcal mol}^{-1} = -12.4 - 6 = -18.4 \, \text{kcal mol}^{-1}. \tag{2.14}$$

It is interesting to consider that the 1000-fold difference in K_A is achieved by only a 29% increase in intrinsic bond energy. That is because the bond energy enters K_A as an exponential.

The primary utility of ΔG_{bond} is that components contributing to ΔG_{bond} are simply additive. This becomes especially important in the calculation of cooperativity, considered in chapter 7.

As a final note, equation (2.9b) may remind the reader of the classical separation of ΔG into an enthalpic and an entropic term:

$$\Delta G_A = \Delta H - T\Delta S$$

where ΔH and ΔS are the changes in enthalpy and entropy upon association. The entropy term is the same as used above if it applies only to the entropy of the protein subunits. However, ΔG_{bond} for protein–protein association is more complicated than a simple ΔH. We will see in chapter 3 that ΔG_{bond} is primarily the free energy from the hydrophobic bond, which is driven by an increase in solvent entropy. This complication means that it is generally not useful to consider enthalpy and total entropy change for protein–protein association. Entropy dominates, with protein subunit entropy favoring dissociation, and a greater solvent entropy favoring association.

Units of energy

We have expressed free energy in the old-fashioned units of kcal mol^{-1}. Physicists today prefer the units kJ mol^{-1} or $k_B T$ (putting it on a single molecule scale). It is useful to know the conversions:

$$1 \text{ kcal mol}^{-1} = 4.184 \text{ kJ mol}^{-1} = 1.689 \, k_B T$$

A useful guide for thinking about energy and forces at the single molecule level is that $k_B T$, which is the average thermal energy buffeting a protein molecule in solution, is equal to

$$k_B T = 4.1 \text{ pN nm (for } T = 298 \, K)$$

For comparison a kinesin motor molecule can generate a force of 5 pN in moving 8 nm on a microtubule. The energy for this motor stroke is 40 $k_B T$.
The intrinsic subunit entropy of 6 kcal mol^{-1} is equal to 10 $k_B T$.

Appendix Using Excel Solver to fit binding data to determine K_D.

Let's assume you did a binding reaction with one fixed concentration of reactant Rtot, and a dozen or so concentrations of reactant Gtot, and that you have assayed for the concentration of the complex $[GR]$ (or converted the measured value of $[G_{free}]$ or $[R_{free}]$ to $[GR]$). The goal is to find the K_D that gives the best fit to the data, using the formula

$$[GR] = \frac{([R_{TOT}] + [G_{TOT}] + K_D) - \sqrt{([R_{TOT}] + [G_{TOT}] + K_D)^2 - 4[R_{TOT}][G_{TOT}]}}{2} \quad (A.1)$$

Install the Solver tool in Excel (select Tools/Excel Add-ins: find the Solver routine and add it. It will then be available on the Tools menu).

Open a new Excel workbook. In cell A2 enter the text label 'Rtot µM.' In A3 insert the concentration of reactant R, which is the same in all measurements. In A5 insert the text label 'Kd fit.' Leave A6 blank, or enter a guess for the value of Kd. Cell A6 will be varied by solver, and the best fit Kd will appear there at the end.

Label column B (text in B1) 'Gtot µM.' In B2–Bn (where Bn is the last data cell) enter the concentration of Gtot in µM for each measurement.

Label column C 'GRmeas µM.' In C2–Cn enter the measured concentration of GR for each concentration of Gtot.

Label column D 'GRpred µM.' In D2 enter the formula:

=((A3 + B2 + A6) − SQRT((A3 + B2 + A6)^2 − 4*A3*B2))/2

Grab the small square lower right of cell D2 and drag down to cover all of the data.

Label column E 'error^2' In cell E2 enter the formula

=(C2 − D2)^2. Drag this down to cover the data.

Now create a new cell in column A. In A8 enter text label 'SUM err^2.' In A9 enter the formula

=SUM(E2:En)

You are now ready to do the fit. Go to Tools/solver. In the solver window 'Set objective' enter A9, and click 'Min.' 'By Changing Variable Cells:' enter A6. 'Select Solving Method': GRC nonlinear. Click solve, and hopefully it will work. If you get a meaningless number in A3 it might help to replace it with a reasonable estimate of Kd, to get the program off to a good start.

The final step is to generate a graph. Select columns B, C, D and Insert/chart/ XYscatter. This will plot columns C and D versus B, giving two curves showing the experimental points and the best fit curve. Click on one curve and see which column is highlighted. Let's say it is C, the GRmeas. You want to keep that as dots only, so you can leave it alone. Click on the other curve, and column D should be highlighted. Here you want to eliminate the dots and show a continuous curve. With that curve highlighted and the cursor on a data point, right click to bring up a menu (select 'format data series' if that appears). Select the paint bucket/line and 'solid line' or 'automatic.' To eliminate the dots select markers/marker options/none.

References

[1] Doty P and Myers G E 1953 Thermodynamics of the association of insulin molecules *Discuss. Faraday Soc.* **13** 51–8

[2] Chothia C and Janin J 1975 Principles of protein–protein recognition *Nature* **256** 705–8

[3] Erickson H P 1989 Cooperativity in protein–protein association: the structure and stability of the actin filament *J. Mol. Biol.* **206** 465–74

[4] Horton N and Lewis M 1992 Calculation of the free energy of association for protein complexes *Prot. Sci.* **1** 169–81

IOP Publishing

Principles of Protein-Protein Association

Harold P Erickson

Chapter 3

The nature of the protein–protein bond, à la Chothia and Janin

In the 1960s and early 1970s x-ray crystal structures were obtained for several protein dimers and oligomers. This permitted one to look at the types of contacts or chemical bonds formed across the interface. The information was staggeringly complex. There were numerous van der Waals (vdW) contacts, hydrogen bonds (H-bonds) and ionic bonds made across the interface. In these early times authors would present tables listing all the vdW, H-bond and ionic contacts made on each side of the interface. Alternatively, a schematic diagram would draw lines between the amino acids making contact across the interface. Figure 3.1 shows such a diagram for the complex of hen egg white lysozyme with an antibody [1, 2], which will be discussed in chapter 4.

Researchers who tried to add up the ΔG_A for all these bonds ran into a contradiction. They used magnitudes of individual bonds taken from physical chemical studies: ionic bonds $\sim$5–10 kcal mol^{-1}, H-bonds $\sim$0.5–5 kcal mol^{-1}, vdW bonds 0.2–0.5 kcal mol^{-1}. If one simply added up all the contributions the total ΔG_A was enormous, much larger than the -14 to -18 kcal mol^{-1} needed for ΔG_{bond} (chapter 2). It was clear that all of these interactions could not be contributing at full strength. Before addressing the resolution of this enigma, we will discuss the nature of H-bonds.

3.1 Hydrogen bonds and ionic bonds in proteins

A hydrogen bond involves a hydrogen atom that is covalently bound to an electronegative atom, like oxygen. The O pulls some negative charge from the H, leaving it with a partial positive charge. The O in turn is left partially electronegative. If this partially charged H approaches another electronegative atom, e.g. a second O, the small positive charge on the H will attract it to the second O. The covalent OH is called the H-bond donor, and the second O the acceptor. The H

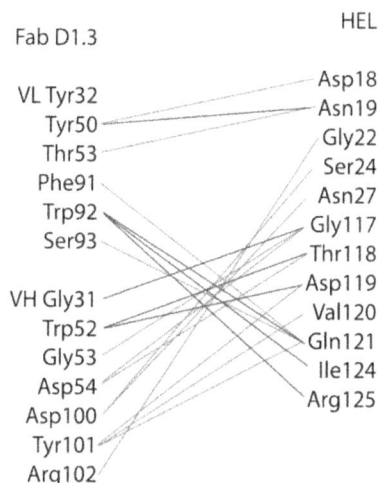

Figure 3.1. redrawn from figure 3 of reference [2]. The two columns list the amino acids of the Fab (left side: VL is the variable light chain domain and VH the variable heavy chain) and HEL (right side) that make contact across the interface. The red lines indicate contacts that form an H-bond, and the blue lines indicate van der Waals contacts.

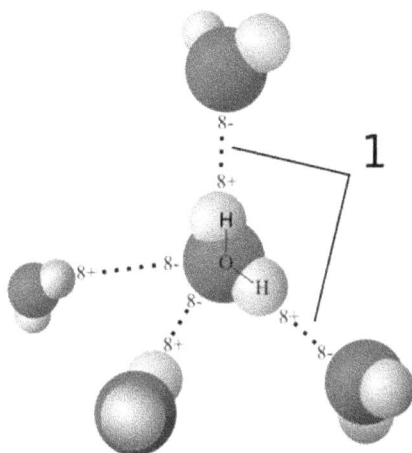

Figure 3.2. Hydrogen bonds between water molecules. This image has been obtained by the author from the Wikimedia website where it was made available under a CC BY-SA 3.0 licence, https://commons.wikimedia. org/w/index.php?curid=14929959. It is included within this article on that basis. It is attributed to Qwerter.

atom is said to be shared by the two electronegative atoms. The H-bond is strongest when the three atoms are roughly linear. Figure 3.2 shows H-bonds between water molecules.

H-bonds can be formed between neutral groups, e.g. the OH of Ser can be a donor bonding to the carbonyl O of the peptide backbone. H-bonds can also form between a charged donor and acceptor. For example, the positively charged primary amine of Lys can form an H-bond to the negatively charged carboxyl of Glu.

Interactions between fully charged groups are called salt bridges or ionic bonds or electrostatic bonds. In every case the bridge contains an H atom, so they can therefore be designated charged H-bonds. Charged H-bonds are more energetic than uncharged H-bonds.

H-bond donors include the peptide backbone and the amino groups of Arg, Lys, Asn, Gln; H-bond acceptors include carbonyl oxygens of the peptide backbone and carboxyl or carbonyl groups of Asp, Glu, Asn, Gln. The O–H of Tyr, Ser and Thr can serve as H-bond donor or acceptor. See figure 3.3 for examples of various H bonds across a protein–peptide interface.

3.2 The simplified protein bond model of Chothia and Janin

In 1975 Cyrus Chothia and Joel Janin published a model of the protein–protein interface that swept away the enormous complexity of ionic bonds, H-bonds and vdW interactions [4]. They basically said you could ignore them for protein–protein bonds made in an aqueous environment. The bond energy is actually generated by a hydrophobic effect that was simply proportional to the area of the interface.

Their insight was partly inspired by a paper the previous year from Fred Richards that described the interior of globular proteins as being close-packed atoms, with few spaces and no water molecules [5]. A crucial observation of Chothia and Janin was that the interface between protein subunits looks like the interior of a protein molecule—closely packed atoms with no spaces or holes for water. This means that

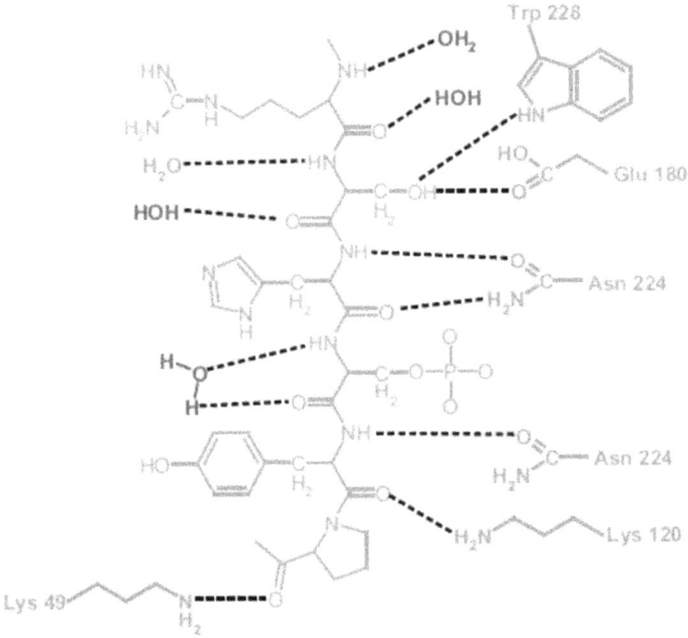

Figure 3.3. Examples of H bonds between a globular protein (green) and a peptide ligand (brown) (reprinted from [3]). Copyright Oldfield *et al*; licensee BioMed Central Ltd. 2008 CC. BY 2.0.

the two surfaces must fit together very snugly, excluding water. This close contact means that there are many vdW contacts across the interface, and no large voids. It was also clear from crystal structures that many H-bonds and ionic bonds were made across the interface. Chothia and Janin then pondered the question of how each of the many interactions would contribute to the free energy of the protein–protein bond. Their great simplification was that the net contribution of ionic, H- and vdW bonds to the ΔG_{bond} was approximately zero.

The key to this analysis was to realize that when the protein–protein bond is broken, the interface residues are not placed in a vacuum, but are exposed to the solvent. The bonds that were formed across the interface when the subunits were together are largely replaced with ones to solvent molecules when the subunits are apart.

Chemical interactions across the protein–protein interface

Number	Type	kcal mol^{-1} in vacuum	kcal mol^{-1} in aqueous solvent
8–13	ionic	5–10	~0
	H-bond	0.5–5	~0
35	van der Waals	~1	~0

What provides the bond energy?
Hydrophobic bond, 0.025 kcal mol^{-1} Å^{-2} buried in the interface.

Figure 3.4 illustrates schematically a protein–protein interface. We should first emphasize that the interface is not a contact of a few atoms, but it covers an extensive area of each subunit. Typically 10%–20% of each protein's surface area

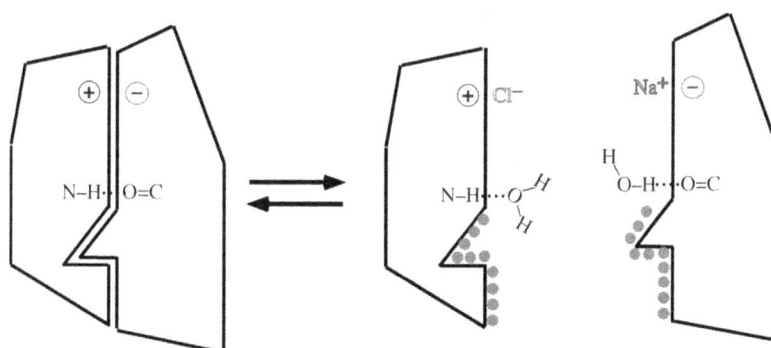

Figure 3.4. Ionic, H- and vdW bonds do not contribute significantly to ΔG_{bond} of the protein–protein interface because they are replaced by equivalent bonds to solvent when the subunits are separated. The small green spheres on the right indicate water molecules replacing the vdW contacts of the interface.

will be in the interface. Second, the interface is very snug, placing the atoms in vdW contact and leaving no holes and no buried water molecules. The beak on the right subunit fitting into the pocket on the left subunit illustrates the snug steric fit. Third, H- and ionic bonds are made across the interface. How much do these contribute to ΔG_{bond}?

The free energy numbers suggested earlier, 1–10 kcal mol^{-1} for H- and ionic bonds, and ~0.5 kcal mol^{-1} for vdW bonds, are for making those bonds in a vacuum. However, protein molecules are in an aqueous solution, and this provides the simplification. Consider first the ionic bond. When the subunits are together, the + charge on one subunit neutralizes the – charge on the other. When the subunits are separated these charges are not bare, but are neutralized by ions in the solution. Similarly for H-bonds: when the subunits are separated both the H-bond donor (N–H) and acceptor (O–C) form H-bonds to water molecules. Chothia and Janin suggested that the ionic and H-bonds to solvent were similar in energy to those across the interface, and thus their net contribution to the intrinsic bond energy, ΔG_{bond}, would be negligible. Similarly, the vdW interactions made across the interface would be replaced with vdW interactions with water. Clearly, the nature of the protein–protein bond is enormously simplified if we simply set all these chemical contributions to zero. What then provides the free energy for protein association?

3.3 The hydrophobic bond

Chothia and Janin concluded that the main source of free energy is the 'hydrophobic bond.' The hydrophobic bond, analyzed by Kaufmann [6], is not actually a bond between hydrophobic residues, but is a complicated effect of solvent entropy. The protein surface is not really hydrophobic—as noted above, water molecules form favorable vdW and H-bonds with the surface. But water molecules have a higher affinity to bond to each other, driven by hydrogen bonding, than to the protein surface. The driving force for the hydrophobic bond comes from the water molecules, and their strong affinity to bond to each other. Figure 3.5 illustrates the basic concept and why it is an entropic effect. In bulk water, away from any edge

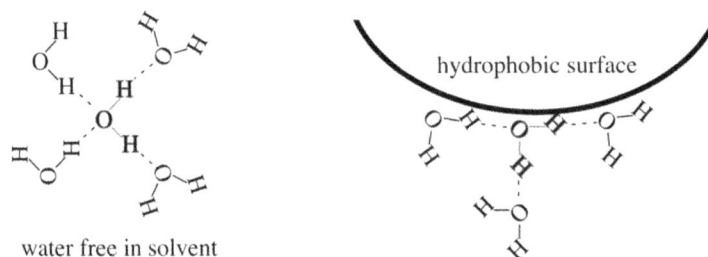

water free in solvent

hydrophobic surface

Figure 3.5. In bulk water each molecule can form about 3.4 H-bonds to neighboring molecules. By breaking and remaking H-bonds the molecules have a high degree of translational and rotational freedom. When a hydrophobic protein surface is introduced, molecules next to it can still make ~3 H-bonds, but they are constrained in their rotational movement.

or surface, each molecule can form a theoretical maximum of 4 hydrogen bonds with neighbors. Liquid water actually averages 3.4 bonds per molecule. In spite of these bonds, in liquid water the molecules remain highly mobile. They are constantly rotating around the single bonds, and also breaking bonds and reforming bonds to new partners. The molecules in the bulk water therefore get the best of both thermodynamic worlds: they achieve close to the maximum enthalpy from H-bonding, and retain a state of high entropy, which is favorable.

At a hydrophobic protein surface, however, there is a constraint. The water molecules next to this surface can't make hydrogen bonds to the hydrophobic amino acids, but they can still make bonds to neighboring water molecules. However, here they have to pay a price—to maximize hydrogen bonds the water molecules must orient themselves in a position that will place their hydrogen bonding sites facing the solvent. The water molecules next to the surface can actually achieve about three hydrogen bonds, so they don't lose much enthalpy. But in order to do this they are much more constrained than those in the bulk solution. These water molecules have less mobility than molecules in the bulk water. They are said to be partially 'frozen.' In thermodynamic terms their entropy is decreased, and this costs free energy.

When the two protein subunits come together to make an interface this layer of 'frozen' water is squeezed out, and released from each surface. The water molecules can now make H-bonds in all directions; they have more freedom, i.e. increased entropy and a lower (more favorable) free energy. It is important to realize that the hydrophobic interaction is not water being repelled by the protein surfaces. The protein–protein association is driven by the increased entropy of the water molecules that are released from the interface surfaces.

Chothia and Janin then argued that the hydrophobic bond energy should be proportional to the amount of water released, which in turn should be proportional to the area of protein surface that is removed from contact with water. Previous studies of model compounds had given a value of 0.025 kcal mol^{-1} for each 1 Å^2 hydrophobic surface removed from contact with water. Chothia and Janin therefore calculated the surface area buried in the interface, which is called the 'water accessible surface area' (WASA). This interface area, including the buried WASA of each subunit, was on the order of 1100–1700 Å^2 for insulin dimer, trypsin-PTI and hemoglobin αβ dimer.

Note that many modern authors use the program PISA to calculate the WASA. This program reports the WASA for one side of the interface, so the total WASA is twice that value. If a reported WASA looks low and the authors don't specify, it is likely that it is one side of the interface, and the total WASA is twice the value.

Using the value of 0.025 kcal mol^{-1} Å^{-2}, Chothia and Janin estimated a hydrophobic free energy of about 30–40 kcal mol^{-1} favoring these associations. These energies are much larger than the 14–18 kcal mol^{-1} intrinsic bond energy for actual protein–protein bonds (chapter 2). This means that the hydrophobic bond energy is more than enough to account for the observed free energy of bonding. The actual ΔG_{bond} is less than the available hydrophobic free energy because realistic proteins have imperfections in the interface that subtract from this maximum (discussed in chapters 6 and 7).

3.4 Complementarity

Earlier we expressed the simplification that the *net* contribution of H-bonds to the protein–protein bond is zero. However, that does not mean that they can be ignored. These bonds play a crucial role in protein association, namely in determining that a protein binds only to its designated partner. Protein association is not promiscuous: e.g. pancreatic trypsin inhibitor associates very strongly with trypsin, but not at all with most other proteins. Interfaces must be highly complementary to achieve this high specificity. According to the Chothia and Janin model this complementarity comprises three features.

a) **Ionic complementarity.** The original formulation of Chothia and Janin suggested that all possible ionic bonds (charged H-bonds) must be made across the interface. If there is a *minus* charge group on one subunit, there must be a *plus* charge opposite on the other side to neutralize it. Otherwise the subunit association would require breaking an ionic bond to solvent (3–6 kcal mol^{-1}) and replacing it with nothing. Burying a charge might be very expensive in free energy and could strongly destabilize the protein–protein interaction. (However, It turns out that ionic complementarity is not nearly as strict as originally thought, as discovered in the 'hot spot' analysis discussed in chapter 6.)

b) **Hydrogen bond complementarity.** The argument for H-bonds is analogous to that for ionic bonds. Any H-bond donor on one subunit must find a H-bond acceptor on the opposite subunit. Losing an uncharged H-bond could cost 0.5–2 kcal mol^{-1}. (Again, however, the 'hot spot' analysis shows that H-bond complementarity is not nearly as strict as originally thought.)

c) **Steric complementarity.** When separated, the subunit interfaces are covered with H$_2$O, and all the atoms on the protein surface make vdW contacts with water molecules. The vdW bonds are less than 1 kcal per atom, but there are lots of atoms. In order not to lose the energy of the vdW interactions, contacts with solvent must be replaced by contacts with the other subunit. This means that the subunits have to fit together very snugly. Separating a pair of atoms even 1 Å will substantially reduce the vdW energy. This is indicated in figure 3.4 by the projecting beak of one subunit fitting into a complementary pocket in the other. (The 'hot spot' analysis showed that steric complementarity is very high over the hot spot, but relaxed somewhat by holes and waters in the peripheral zone. However, a protruding amino acid on one interface must find a pocked on the other. Steric complementarity is the major determinant of the specificity of protein–protein association.)

Hydrophobic and hydrophilic amino acids

Amino acids that are non-polar are considered hydrophobic. Large hydrophobic amino acids are Trp, Phe, Pro, Ile and Leu. Smaller hydrophobic amino acids are Cys, Val, Ala, Gly. Polar amino acids are ones whose side chains can make H-bonds. Uncharged polar amino acids are Asn, Gln, Ser, Thr and Tyr. The indole N of Trp can also make an H bond, but its large aromatic character dominates. Charged polar amino acids are Asp, Glu, Arg, Lys, and His (His may have a + charge or be neutral). But note that some of these polar and charged amino acids have large hydrophobic segments, e.g. the aliphatic chains of Lys and Arg, and the aromatic ring of Tyr. These can support substantial hydrophobic interaction as well as a H-bond. Hydrophobic amino acids predominate in the interior of globular proteins and to a lesser extent in interfaces.

The Chothia–Janin treatment of interfaces, however, considers hydrophobic and polar amino acids to be the equivalent, because it is assumed that all H-bonds are made across the interface. In this case the only thing that matters is the surface area.

Janin and Chothia [7] summarized the structure of the protein–protein bond: 'The protein–protein recognition sites discussed here (protease–protease inhibitors and antigen–antibody) have very similar structural properties: 34 ± 7 close-packed contact residues bury 1600 ± 350 $\approx^2$ of surface (total from two sides of the interface) and form 8–13 intermolecular hydrogen bonds. The conformational changes that occur on association involve only the rotation of certain side chains and small movements in some sections of the main chain.'

3.5 Final comments: what is the mechanical rigidity of globular proteins and their polymers?

Gittes and Howard [8] addressed this question by measuring thermal fluctuations of actin and microtubules. They reported that:

'The microtubule's Young's modulus is ~1.2 GPa, similar to Plexiglas and rigid plastics.'

Later studies have debated this and reported lower values, suggesting that microtubules were as flexible as Teflon. However, Joe Howard has found flaws in these studies, and his latest analysis strongly supports the Plexiglas rigidity [9].

This Young's modulus should apply to globular protein molecules in general and to their polymers—a very important concept. Proteins and their polymers are not 'soft wet materials' as some engineers present them, but are quite rigid. They are mechanically equivalent to the hardest plastics. This suits them nicely to make the machinery of life.

A related question is: how much can you bend the protein subunits about the bond interface? Not very much. The left diagram in figure 3.6 shows the correctly docked interface, and right diagram shows a slightly bent interface. If the tilt pulls

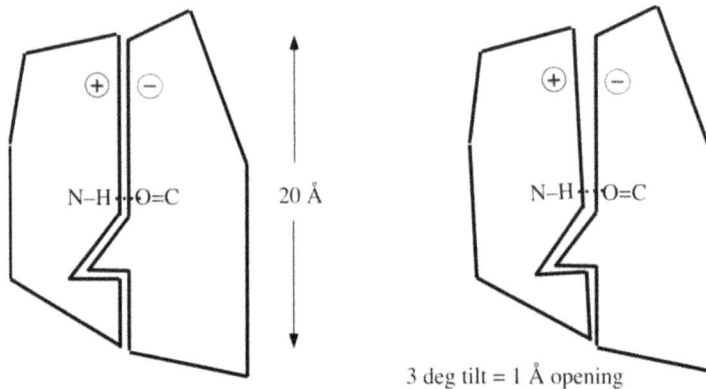

Figure 3.6. A tilt of 3° will produce a 1 Å opening across a 20 Å interface. A larger tilt will strain bonds and let in water.

the subunits apart by about 1 Å, it will strain but not completely disrupt van der Waals and hydrogen bonds. Opening it any more would let water into the hydrophobic interface. It is a reasonable guess that this 1 Å split would be the maximum before the bond is disrupted. Across the 20 Å interface this would mean a 2.8° tilt.

The take-home of this brief analysis is that protein subunits could rock back and forth by only about 3° and remain attached. Rotation around the plane of the interface would also be highly restricted, also about 3°. The protein–protein bond is quite rigid in all directions.

References

[1] Amit A G, Mariuzza R A, Phillips S E and Poljak R J 1986 Three-dimensional structure of an antigen–antibody complex at 2.8 A resolution *Science* **233** 747–53

[2] Fischmann T O, Bentley G A, Bhat T N, Boulot G, Mariuzza R A, Phillips S E, Tello D and Poljak R J 1991 Crystallographic refinement of the three-dimensional structure of the FabD1.3-lysozyme complex at 2.5-Å resolution *J. Biol. Chem.* **266** 12915–20

[3] Oldfield C J, Meng J, Yang J Y, Yang M Q, Uversky V N and Dunker A K 2008 Flexible nets: disorder and induced fit in the associations of p53 and 14-3-3 with their partners *BMC Genomics* **9** S1

[4] Chothia C and Janin J 1975 Principles of protein–protein recognition *Nature* **256** 705–8

[5] Richards F M 1974 The interpretation of protein structures: total volume, group volume distributions and packing density *J. Mol. Biol.* **82** 1–14

[6] Kauzmann W 1959 Some factors in the interpretation of protein denaturation *Adv. Protein Chem.* **14** 1–63

[7] Janin J and Chothia C 1990 The structure of protein–protein recognition sites *J. Biol. Chem.* **265** 16027–30

[8] Gittes F, Mickey B, Nettleton J and Howard J 1993 Flexural rigidity of microtubules and actin filaments measured from thermal fluctuations in shape *J. Cell. Biol.* **120** 923–34

[9] Howard J 2008 Molecular Mechanics of Cells and Tissues *Cell. Mol. Bioeng.* **1** 24–32

IOP Publishing

Principles of Protein–Protein Association

Harold P Erickson

Chapter 4

The structure of an antibody bound to its protein ligand—lock and key versus induced fit and conformational selection

Directed reading:

Amit A G, Mariuzza R A, Phillips S E and Poljak R J 1986 Three-dimensional structure of an antigen–antibody complex at 2.8 A resolution *Science* **233** 747–53 [1].

This paper [1] reports the first crystal structure of an antibody bound to a protein antigen. The interface corresponds very well to the simple model of Chothia and Janin [2], and is interpreted largely in their terms. We will provide some background and then discuss some highlights of the paper. The reader is urged to read the original paper in detail.

The antibody used here was a Fab fragment of the monoclonal antibody D1.3, against hen egg white lysozyme (HEL). The large, bivalent IgG antibody can be cleaved by proteases to produce one Fc fragment and two Fab fragments. Each Fab contains four Ig domains and binds one HEL. The structure of the Fab–HEL complex is shown in figure 4.1 using the modern cartoon format, which is clearer than the alpha carbon stick diagram in figure 1 of Amit *et al*. Figure 4.2 shows a spacefill model of the key contacts made by Gln121.

The authors first addressed a controversy in protein–protein association that was especially relevant to antibodies and their protein antigens. The Chothia–Janin model postulated that the protein subunits were rigid bodies, preformed to fit together to make the highly specific interface. By 1986 this had been confirmed for quite a few protein–protein complexes, and was referred to as the 'lock and key' model. However, the antibody field faced a problem—how to explain the vast repertoire of antibodies that could bind virtually any protein ligand. One possibility was an 'induced fit' model, where the Fab would have a somewhat flexible structure that could mold itself to fit a range of different antigens. The induced fit model suggested also that the antigen might be altered to fit the Fab. This is in striking contrast to the rigid, preformed structures of the lock and key.

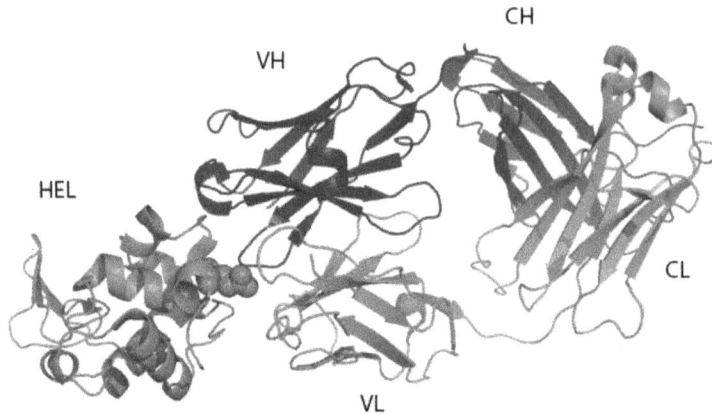

Figure 4.1. Cartoon diagram of the HEL–D1.3 Fab complex. HEL is in red. The Fab comprises four Ig domains, VH and CH from the heavy chain (blue) and VL and CL from the light chain (green). V stands for variable: the VH and VL domains have three hypervariable loops (complement determining regions, CDR1–3 in Amit *et al*) that face the HEL and form the specific protein–protein interface. Produced from pdb 1FDL [1], and displayed in PyMol (The PyMOL Molecular Graphics System, Version 2.0 Schrödinger, LLC).

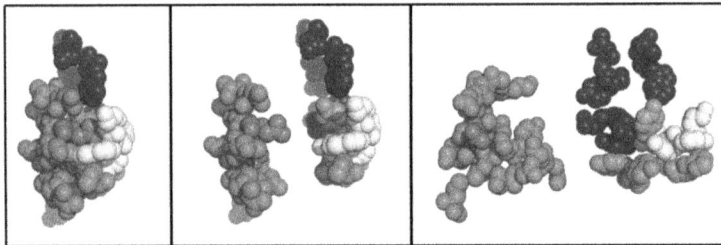

Figure 4.2. Spacefilling model of the amino acids making contact in the HEL–D1.3 Fab complex. HEL is red and VH and VL are blue and green, as in figure 4.1. Gln 121 is purple, and the Fab amino acids contacting Gln 121 are cyan for VH and yellow for VL. The left panel shows the interface together. The middle panel shows the HEL and Fab pulled apart. The right panel shows the interfaces rotated 90 degrees. Produced from pdb 1FDL [1], and displayed in PyMol (The PyMOL Molecular Graphics System, Version 2.0 Schrödinger, LLC).

Amit *et al* came down firmly on the side of lock and key, at least for the HEL protein. The structure of HEL alone had already been determined in several crystals. Superimposing the Cα chains of the isolated HEL on that of the Fab complex showed rms differences of ~0.6 Å. This is equivalent to the noise in any given crystal structure, and to the differences between different crystal structures of the same protein. Amit *et al* concluded 'Thus, complex formation with antibody D1.3 produces no more distortion of the structure of lysozyme than does crystallization.'

At the time of this article there was no separate crystal structure of the Fab D1.3. A later study obtained a structure of the free Fab and it showed movements of 0.5–0.7 Å relative to the Fab in the complex [3]. These small movements are similar to those observed for HEL monomer versus the complex, and fits the 'lock and key' model.

The authors' description of the interface is a close match to the idealization of Chothia and Janin. 'The interface between antigen and antibody extends over a large area with maximum dimensions of about 30 by 20 Å. The antibody combining site appears as an irregular, rather flat surface with protuberances and depressions ... The interacting surfaces are complementary, with protruding side chains of one lying in depressions of the other, in common with other known protein–protein interactions [2]. There are many van der Waals interactions interspersed with hydrogen bonds ... Many hydrogen bonds occur between the side chains of the antigen and the main polypeptide chain of the antibody, and vice versa ... In all, 748 $Å^2$ or about 11 percent of the solvent-accessible surface of lysozyme is buried on complex formation, together with 690 $Å^2$ for the antibody.'

Gln 121 is singled out as very important. 'There is a small cleft between the third CDR's of VH and VL ... [that] accepts the side chain Gln 121 of lysozyme.' The Gln 121 would correspond to the beak in figure 3.3, and the cleft to the pocket that accepts it. In addition to the steric fit, 'its amide nitrogen forms a strong, buried hydrogen bond to the main chain carbonyl oxygen of VL Phe 91.'

4.1 Nature's site-directed mutagenesis experiment

Today we would test the importance of Gln 121 by site-directed mutagenesis, changing it to Ala and seeing how that affected the binding affinity. The molecular biology to do this did not exist in 1986, but Amit *et al* found the equivalent experiment done by Nature.

In 1986 the lysozymes from a number of bird species had been sequenced. Most differed from hen lysozyme at only a few amino acid positions. The Poljak group had determined that Fab D1.3 bound HEL with $K_A = 4.5 \times 10^7$ M^{-1}, which is typical for monoclonal antibodies binding to protein antigens (range 10^5–10^{10} M^{-1}). They collected lysozymes from different bird species and tested their affinity for Fab D1.3. 'Bobwhite quail lysozyme, with four amino acid sequence differences from hen lysozyme, but none in the interface with Fab D1.3, binds with similar affinity. The binding to lysozymes of partridge ... California quail ... Japanese quail, turkey ... pheasant and guinea fowl is undetectable ($K_A < 10^5$ M^{-1}) ... These lysozymes differ from hen lysozyme at the amino acid residue at position 121.'

Why is Gln 121 so important? As noted above, its nitrogen makes a strong H-bond to a backbone carbonyl on VH. In addition, its side chain makes hydrophobic contacts with Tyr101, Tyr 32 and Trp 92 of VL (figure 3.2). We will see later that this Gln 121 fits the definition of a classic hot spot residue, i.e. a residue whose mutation substantially weakens the interface.

The paper of Amit *et al* also described features of general importance for understanding antibody binding to protein ligands. 'The lysozyme antigenic determinants recognized by D1.3 are made up of two stretches of polypeptide chain, comprising residues 18 to 27 and 116 to 129, distant in the amino acid sequence but adjacent on the protein surface.' This makes the important point that a protein antigenic determinant will typically include more than a single peptide. 'All six CDR's interact with the antigen and in all, 16 antigen residues make close contacts with 17 antibody residues.' These contacts are shown schematically in figure 3.1, as refined from a later higher resolution structure [4]. Some contacts involve only vdW contacts, but most involve H-bonds. Overall, the picture is an extensive interface involving many vdW contacts and H-bonds across the interface.

4.2 Induced fit and conformational selection

Although the Poljak group initially championed a lock and key model for the D1.3–HEL study [1], they later obtained evidence for some movement in the antibody that they interpreted as induced fit [3]. However, the movements were only 0.5–0.7 Å. Larger movements were later found for some antibody–antigen complexes and were interpreted as induced fit. However, it now seems that most of these can be attributed to conformational selection.

The question of induced fit versus conformational selection was comprehensively addressed by Stein *et al* [5], who compared the structures of 2090 proteins in unbound versus bound states. They found that 65% of the pairs showed rmsd movements <1.5 Å, which they considered to be within normal thermal movements. This 65% of protein interaction pairs are considered lock and key. For pairs with larger movements they considered two possibilities. First, conformational selection could occur when one of the proteins had inherent flexibility that produced a range of conformations. The binding partner could select and lock in place one of these conformations. Second, a true induced fit would occur when the final structure was outside the normal range of flexible movements. To distinguish between these they applied computer modeling to estimate the range of flexible movements in the monomer. Their criteria placed only 0.8% of protein complexes in the induced fit category, while 38% of all protein complexes were classified as conformational selection (or an ambiguous twilight zone) where the structure in the complex was selected from the range of normal motions of the unbound protein.

In summary, about two-thirds of protein complexes fit the lock and key model, with only small movements upon complex formation. Almost all of the structures showing larger movements can be classed as conformational selection, where a flexible loop assumes multiple conformations, one of which is selected upon binding. True induced fit occurs but is rare.

References

[1] Amit A G, Mariuzza R A, Phillips S E and Poljak R J 1986 Three-dimensional structure of an antigen–antibody complex at 2.8 A resolution *Science* **233** 747–53
[2] Chothia C and Janin J 1975 Principles of protein–protein recognition *Nature* **256** 705–8
[3] Bhat T N, Bentley G A, Fischmann T O, Boulot G and Poljak R J 1990 Small rearrangements in structures of Fv and Fab fragments of antibody D1.3 on antigen binding *Nature* **347** 483–85
[4] Fischmann T O, Bentley G A, Bhat T N, Boulot G, Mariuzza R A, Phillips S E, Tello D and Poljak R J 1991 Crystallographic refinement of the three-dimensional structure of the FabD1.3-lysozyme complex at 2.5-Å resolution *J. Biol. Chem.* **266** 12915–20
[5] Stein A, Rueda M, Panjkovich A, Orozco M and Aloy P 2011 A systematic study of the energetics involved in structural changes upon association and connectivity in protein interaction networks *Structure* **19** 881–89

IOP Publishing

Principles of Protein–Protein Association

Harold P Erickson

Chapter 5

The complex of growth hormone with its receptor—one protein interface binds two partners

Directed reading:

DeVos A M, Ultsch M and Kossiakoff A A 1992 Human growth hormone and extracellular domain of its receptor: crystal structure of the complex *Science* **255** 306–12 [1].

In chapter 3 we emphasized the role of complementarity in determining the exquisite specificity of a protein for binding only one protein partner. Steric complementarity required that the two interfaces fit snugly together. Ionic and H-bond complementarity required that these H-bonds were completed across the interface, so that no unpaired donors or acceptors were buried. All of this seemed to ensure that a protein would only form a bond to its single designed partner. For a great majority of protein–protein pairs this is true; however, some proteins have an interface that can form high affinity complexes with two or more binding partners.

5.1 GHR binds two different patches on opposite sides of GH

Many receptors with single-transmembrane domains operate through a mechanism of receptor dimerization. For example, the hormone PDGF is a dimer, which binds the extracellular domains of two PDGF receptors. When the two receptors are thus brought together, their cytoplasmic kinase domains phosphorylate each other and initiate a signaling cascade.

An intriguing twist on receptor dimerization was discovered for the complex of growth hormone (GH) and its receptor (GHR). It turns out that monomeric GH binds two GHR, bringing the cytoplasmic domains together to initiate a signaling cascade. How can one asymmetric GH bind two GHRs? Obviously GH must have two, non-overlapping patches that can each bind a GHR. The GH monomer shows no sign of symmetry, so these two binding sites on GH must be different. One possibility was that each patch on GH binds to a completely different patch on GHR. However, the crystal structure showed a surprising mechanism—the two different patches on GH bind the essentially same patch on the two GHRs.

The structure of GH bound to two GHRs is shown in figure 5.1. GH in the complex is an alpha helical bundle identical to the structure previously determined for GH alone. The GHRs have a simple structure comprising two beta sandwich domains connected by a short linker. This beta sandwich structure is now known as the FN-III domain, one of the most abundant domains in the protein database (discussed in chapter 10). The bend between the domains is the same in both GHRs, suggesting that it is a rigid feature of the receptor. Both GH patches bind at the bend, and both FN-III domains of GHR contribute to each interface.

The binding sites of the two GHRs on opposite sides of the single GH are designated sites 1 and 2. The interfaces on GH are quite different. Site 1 on GH is somewhat concave, while site 2 is flat. The two receptors have identical structures, with an rms difference in Cα of 1.0 Å; i.e. there is no induced-fit movement. Remarkably, each of the GHRs use essentially the same patch to bind the two different patches on GH. The site 1 patch is larger than site 2: the WASA is 2460 Å^2 for site 1 and 1800 Å^2 for site 2 (de Vos *et al* reported the WASA for one side; I have

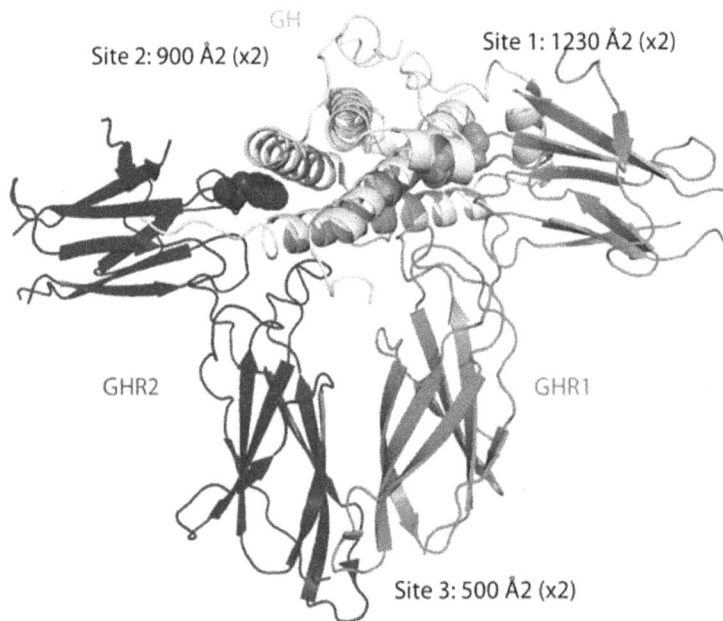

Figure 5.1. The GH–GHR complex from pdb-3HHR [1]. GH (yellow) is a four-helix bundle. GHR1 (green) and GHR2 (blue) each consist of two FNIII domains (a beta sandwich with 3 strands on one side and 4 strands on the other). The important W104 is shown in spacefill. Sites 1 and 2 are the interfaces of GHR1 and GHR2 on opposite sides of GH. Site 3 is the interface of the C-terminal FNIII domains of the two receptors with each other. A C-terminal 8-amino-acid peptide, not resolved, provides a flexible link from the bottom of the C-terminal domains to the membrane. Displayed in PyMol.

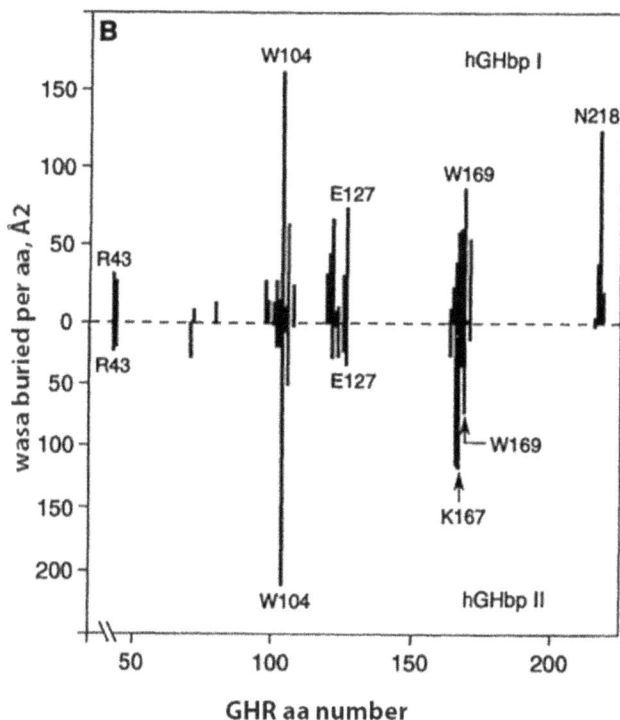

Figure 5.2. Buried WASA calculated for each GHR for binding to GH site 1 (top) and site 2 (bottom). Note that mostly the same GHR amino acids contribute to each binding site. Reprinted from [1] with permission from AAAS.

doubled it to report the total). It should be noted that 2460 $Å^2$ WASA for site 1, which has $\sim$ 1.1 nM KD, is substantially larger than the $\sim$1600 $Å^2$ for most previously reported high affinity interfaces. This interface is apparently less efficient than others, with defects in the complementarity. We will return to this in the next chapter on the 'hot spot.'

De Vos *et al* extended the WASA calculation down to single amino acids, calculating how much area of each amino acid of GHR was buried in the interface. Data shown in figure 5.2 makes the important point that the same GHR amino acids are buried in each interface. W104 buries the largest area, 160 $Å^2$ in site 1 and 210 $Å^2$ in site 2. W169 and adjacent amino acids also bury significant WASA. R43 contributes buried WASA and also H-bonds to each site. N218 is an outlier: it contributes substantially to site 1 but not at all to site 2. In addition to a larger WASA, site 1 has nine H-bonds, compared to four for site 2. All of this is consistent with site 1 being substantially higher affinity than site 2.

Cunningham *et al* [2] had previously obtained evidence that the binding of GHR2 to site 2 could only occur after GHR1 had bound to site 1. The crystal structure shows how this works. The buried WASA of site 2 is 27% less than that of site 1. If the sites are equally efficient in utilizing the WASA, ΔG_{bond} of site 2 would be 27% less than that of site 2. This would weaken the K_A of binding by several orders of magnitude (see equations (2.11)–(2.14) in chapter 2). However, the crystal structure shows that when GHR2 binds site 2, it also brings the C-terminal FN-III domains together at site 3. Site 3 buries 500 $\mathring{A}^2$ WASA, so the total for sites 2 and 3 is 2800 $\mathring{A}^2$. The total WASA for sites 2 and 3 is somewhat larger than the 2460 $\mathring{A}^2$ of site 1. This calculation will be pursued in chapter 7 'Cooperativity.'

One might imagine that the GHR interfaces to sites 1 and 2 could involve substantial conformational changes, i.e. an induced fit; however, their structures showed an rms difference in Cα of 1.0 $\mathring{A}$, consistent with a rigid structure and a lock and key mechanism. De Vos *et al* did note some small movements. 'The difference in Cα position of Trp104 is 2.8 $\mathring{A}$, and the side chain orientation differs in the two receptors. Loop 163–168 also takes on a different conformation,' with Cα movements of 2–4 $\mathring{A}$. These are likely within the ensemble of motions of the free monomer, and therefore an example of conformational selection.

The above analyses were specifically for the soluble ectodomain of GHR, formally designated 'GH binding protein.' Recent work has shown that the transmembrane receptor exists as a weak dimer. The binding of GH causes a large rearrangement of the ectodomains, which is transmitted to the cytoplasmic domains by a scissoring movement to initiate the JAK2 signaling cascade [3].

5.2 Other proteins with multiple binding partners

Although it was surprising to see the same patch on GHR binding two completely different patches on GH, this kind of multiple binding partners has been seen for some other proteins. Delano *et al* [4] explored the Fc fragment of human IgG, which is known to bind at least four unrelated proteins. All four proteins bind to the same site on Fc, a small patch of 1480 $\mathring{A}^2$ WASA (figure 5.3 shows two of these). DeLano *et al* then created a large library of cyclic peptides and selected for binding to Fc.

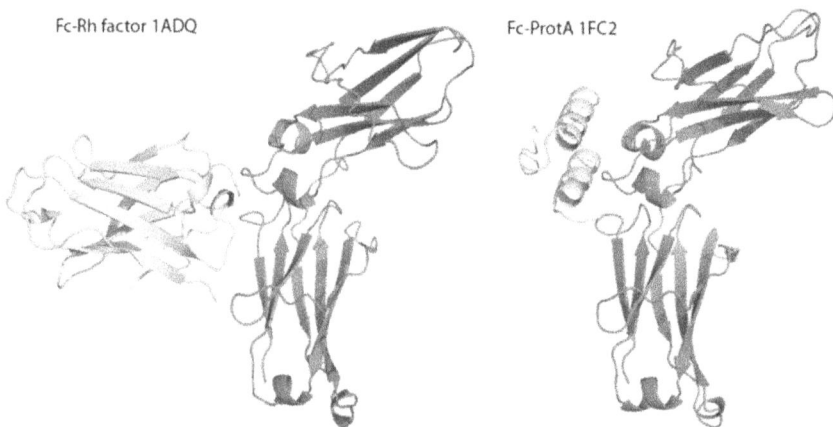

Fc-Rh factor 1ADQ   Fc-ProtA 1FC2

Figure 5.3. The Fc fragment of human IgG binds to at least four proteins. The crystal structures are shown here for rheumatoid factor [5] and Protein A [6]. Displayed in PyMol (The PyMOL Molecular Graphics System, Version 2.0 Schrödinger, LLC).

In principle the peptides could bind anywhere on the surface of the Fc, but the two highest affinity peptides were bound to precisely the same small patch as the four proteins. This strongly suggests that this patch has properties that favor or enable binding to different partners. Interestingly, the Fc protein looks remarkably like the GHR. The Ig domains of Fc are 7-strand β sandwiches very similar to the FN-III domains of GHR. Both Ig and GHR have a bend between the two domains, and the ligands bind at this bend.

The crystal structures of the complexes of Fc with the four proteins and the peptides suggested several features of the Fc patch that might facilitate multiple binding partners. Most important, there were relatively few hydrophilic amino acids in the patch; this left the binding patch primarily hydrophobic. Also, some side chains in the Fc patch underwent substantial movements to accommodate side chains of the different binding partners. These movements would fit under the scheme of conformational selection, since they are energetically feasible for the unbound protein.

A protein patch that can accommodate two or more binding partners should not be considered promiscuous. These binding partners have co-evolved with the target protein to achieve the dual binding. GHR and Fc still retain a very high degree of specificity, binding only the two or so partners from among the thousands of proteins that they can encounter.

References

[1] DeVos A M, Ultsch M and Kossiakoff A A 1992 Human growth hormone and extracellular domain of its receptor: crystal structure of the complex *Science* **255** 306–12

[2] Cunningham B C, Ultsch M, DeVos A M, Mulkerrin M G, Clauser K R and Wells J A 1991 Dimerization of the extracellular domain of the human growth hormone receptor by a single hormone molecule *Science* **254** 821–5

[3] Brooks A J, Dai W, O'Mara M L, Abankwa D, Chhabra Y, Pelekanos R A and Waters M J 2014 Mechanism of activation of protein kinase JAK2 by the growth hormone receptor *Science* **344** 1249783

[4] DeLano W L *et al* 2000 Convergent solutions to binding at a protein–protein interface *Science* **287** 1279–83

[5] Corper A L, Sohi M K, Bonagura V R, Steinitz M, Jefferis R, Feinstein A, Beale D, Taussig M J and Sutton B J 1997 Structure of human IgM rheumatoid factor Fab bound to its autoantigen IgG Fc reveals a novel topology of antibody–antigen interaction *Nat. Struct. Biol.* **4** 375–81

[6] Deisenhofer J 1981 Crystallographic refinement and atomic models of a human Fc fragment and its complex with fragment B of protein A from *Staphylococcus aureus* at 2.9- and 2.8-Å resolution *Biochemistry* **20** 2361–70

Chapter 6

The hot spot in protein–protein interfaces

Directed reading:

Clackson T and Wells J A 1995 A hot spot of binding energy in a hormone–receptor interface *Science* **267** 383–6 [1].

In the simplest Chothia and Janin model (chapter 3) the free energy of the protein–protein bond (ΔG_{bond}) was proportional to the area of the interface, with a value of 0.025 kcal mol^{-1} Å^{-2} buried in the interface. One problem was that the ΔG_{bond} predicted from this area was up to twice the experimentally determined value. I suggested in chapter 3 that the excess might be accounted for by imperfections in the interface: e.g. H-bonds with imperfect geometry, or non-optimal vdW contacts. I did not address how these imperfections might be distributed, but by default we might have expected a uniform distribution. The hot spot papers showed that this assumption was wrong.

To address this question, the group of James Wells capitalized on their crystal structure of growth hormone (GH) bound to its receptor (GHR, called GHbp, for GH binding protein, in the Clackson and Wells paper). In this ambitious study they mutated to Ala, one at a time, every amino acid that made contact across the interface, and determined how the mutation affected the binding affinity. This is called 'alanine scanning.' The mutation to Ala removes all of the side chain above the Cβ, eliminating all vdW contacts and any H-bonds bonds that particular amino acid contributed to the interface. (Mutation to Ala is preferred to Gly because Gly can permit additional conformations of the Cα chain.) This should determine what each amino acid side chain was contributing to the bond energy, ΔG_{bond}.

According to the simple Chothia and Janin model we would expect that deleting an H-bond donor or acceptor, especially if charged, would be costly, since it would leave its partner unpaired. Similarly, removing a large hydrophobic side chain would leave a hole lacking vdW contacts. We would therefore expect most mutations to Ala to weaken the bond. However, this was not the case. Clackson and Wells found that only about 1/3 of the amino acids making contact contributed significantly to the bond energy. 2/3 of the amino acids making contact showed minimal change in binding when mutated to Ala. They also found that the critical amino acids could not be predicted based on H-bonds and buried WASA.

6.1 Hot spot paper one—the technology and alanine scanning of GH

The paper presented for reading, alanine scanning of GHR, is the second in the series [1]. That paper was selected for reading because of its brevity and because it completed the story. It was preceded by alanine scanning of GH [2]. This much longer paper describes the technology in detail and presents the first understanding of the 'hot spot.' In this first paper the hot spot was referred to as the 'functional epitope,' a term that is descriptive but not flashy. We will discuss this paper first.

The key technology was to use plasmon resonance (a Biacore instrument) to measure the kinetic constants k_2 and k_{-1} of GH binding to GHR at site 1. The GHR was coupled to the dextran support using a clever method that eliminated binding to site 3 and consequently to site 2 (see the paper for details [2]). GH or the GH mutant was flowed through the chamber, followed by a wash, to determine the two kinetic constants, and from these to determine K_D and ΔG. The important measure was how much the Ala mutant changed ΔG: this change in binding free energy is termed $\Delta\Delta G$—see box 6.1 for the definition.

The project undertaken by Cunningham and Wells [2] was ambitious in scope for that time. They mutated to Ala each of the 31 amino acids of GH that made contact with the GHR, and measured the binding affinity to GHR. Interestingly, all but one of the mutant proteins could be produced as a soluble protein. This is consistent with the idea that surface amino acids can generally be mutated without compromising the folding of a globular protein. The $\Delta\Delta G$s ranged from a very destabilizing +2.4 kcal mol^{-1}, to –0.9 kcal mol^{-1} (the negative $\Delta\Delta G$ means that removal of that side chain actually enhanced the binding). Overall, there were a number of very surprising conclusions relative to expectations from the Chothia–Janin model.

Box 6.1. Definition of $\Delta\Delta G$, the change in free energy caused by a mutation.

ΔGwt $= -RT\ln K_A$wt $= +RT\ln K_D$wt
ΔGmut $= -RT\ln K_A$mut $= +RT\ln K_D$mut (here the mutation can be on either GH or GHR)
$\Delta\Delta G = -\Delta G$wt $+ \Delta G$mut
A positive $\Delta\Delta G$ means that the mutant binds more weakly. In some cases the mutant bound more tightly than wild type, giving a negative $\Delta\Delta G$. Note that these papers did not separate out a ΔG_{bond}. This was not necessary: the $\Delta\Delta G_{bond}$ would be the same as $\Delta\Delta G$, because ΔG_S is the same 6 kcal mol^{-1} for wild type and mutant.

- 'Of the 31 buried side chains, only eight are needed to account for 85% of the total change in binding free energy resulting from the alanine substitutions.' These had $\Delta\Delta G$ ranging from +1.1 to +2.4 kcal mol^{-1}, corresponding to a 7- to 60-fold increase in K_D (weaker binding affinity).

- 'Eleven side-chains have essentially no effect on overall affinity (each causing less than a 2-fold reduction in affinity). Five side-chains actually hinder binding because when they are converted to alanine we see enhancements in affinity of 2- to 6-fold.'
- There was almost no correlation of $\Delta\Delta G$ with the buried WASA of the side chain, nor with the number of vdW contacts it made or its H-bond formation. 'Thus, while buried surface area, number of vdW contacts ... are useful correlates for general binding affinity, we find that these are poor predictors of the role of individual side chains in this epitope.'
- A very important observation was confirmed in the next paper and led to the term 'hot spot.' 'We find the residues important for binding cluster in a small region near the center of the structural epitope. The functionally 'null' contact residues tend to be near the periphery.'

6.2 Hot spot paper two—scanning GHR and matching the hot spots

The first study by Cunningham and Wells [2] did the comprehensive alanine scanning on the GH side of the complex. The follow-up by Clackson and Wells [1] scanned the GHR side of the complex: they mutated each of the 33 contact amino acids of the GHR site 1 to Ala. As in the previous study they found that, 'fewer than half of the mutations caused substantial loss in binding affinity' (11 hot spot amino acids out of 33 total contact amino acids). In this case two amino acids were hugely important for binding: W104 and W169, where the affinity of the mutant was too weak to measure ($\Delta\Delta G > 4.5$ kcal mol^{-1}). This is reminiscent of the Gln 121 of HEL (hen egg-white lysozyme), where mutation to any other amino acid reduced to immeasurable its affinity to the antibody D1.3 ([3] and chapter 4). This is also expected from the Chothia–Janin model, since Trp and Gln bury large surface areas.

Figure 6.1 shows the main result of this study. The amino acids that contribute most to binding affinity are clustered in the middle of the contact interface. Clackson and Wells called this cluster the 'hot spot.' The hot spot amino acids are almost all hydrophobic. The hot spot is surrounded by amino acids that are part of the contact surface as seen in the x-ray structure, but that don't contribute much to $\Delta\Delta G$. These peripheral, non-hot spot, amino acids are mostly polar or charged.

Clackson and Wells then compared their hot spot on GHR to that previously determined on GH and made the important conclusion: 'Now that both sides of the interface have been mutated systematically, the two functional epitopes can be compared (figure 6.2). This reveals a striking complementarity—the energetically critical and unimportant regions on one molecule match those on the other. Most of the important residues on GH are involved in forming a hydrophobic pocket that closely docks the side chains of Wl04 and W169 from the GHR.'

Why are the peripheral amino acids so unimportant for binding affinity? An important clue was that the peripheral areas appeared to be poorly packed, with gaps that were often filled with well-ordered water molecules. Recall that water is largely excluded from the classical Chothia–Janin interface. Importantly, the bound waters in the peripheral region frequently participated in H-bond networks to the

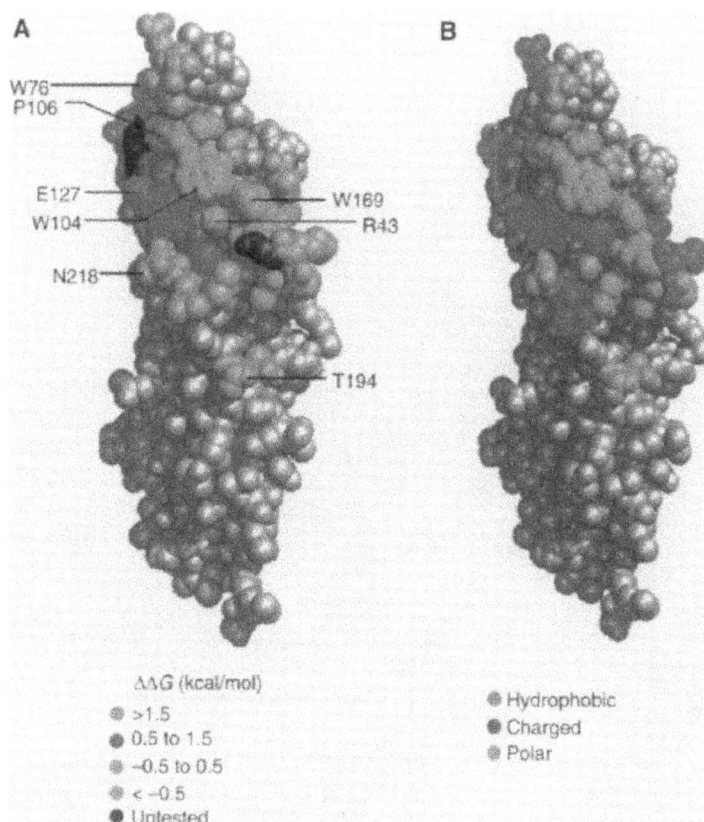

Figure 6.1. (figure 2 in C–W). (a) shows the location of contact amino acids colored according to their $\Delta\Delta G$. Hot spot amino acids ($\Delta\Delta G > 1.5$ kcal mol^{-1}) are red. Importantly the hot spot amino acids are clustered together at the center of the contact interface, surrounded by amino acids that contribute less to $\Delta\Delta G$. (b) colors the amino acids according to their hydrophobicity. The hot spot amino acids are mostly hydrophobic (red), while the peripheral amino acids are polar or charged. Reprinted from [1] with permission from AAAS.

amino acids on each side. This can explain why eliminating one side of an H-bond can have minimal effect on $\Delta\Delta G$. The network of H-bonds across the interface has a plasticity, especially where the bound waters participate, and H-bonds can rearrange to compensate.

6.3 Plasticity in the evolution of protein–protein interfaces

Plasticity of the protein–protein interface was emphasized in a subsequent study from the Wells lab. Atwell *et al* [4] started with the W104A mutant of GHR, whose affinity to GH was too weak to measure, and created random mutations in five amino acids of GH that made contact with W104. They selected for mutants that restored binding, and found one that restored binding to 14 nM (vs 0.3 nM for the wt GH). A crystal structure of this mutant complex showed multiple rearrangements. The cavity left by removing W104 was largely filled by a new Tyr on GH, and

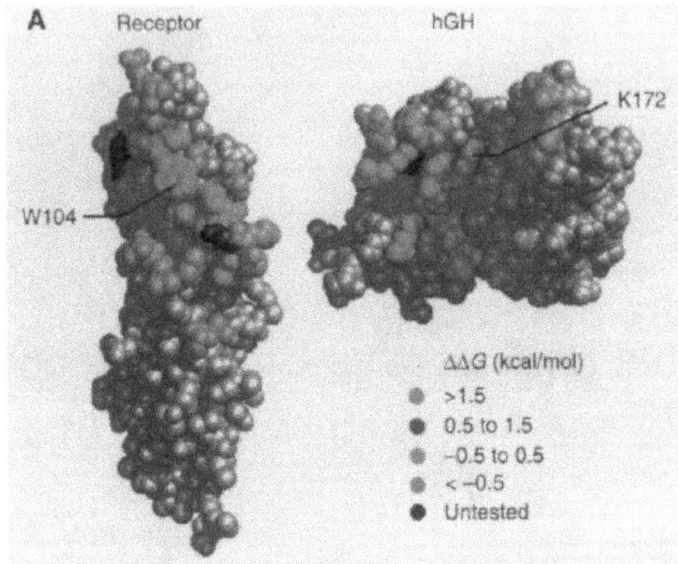

Figure 6.2. (figure 3A in C–W). The GHR–GH complex has been separated and rotated so the contact surfaces face outward. Amino acids are color coded according to their $\Delta\Delta G$. The red amino acids caused the most substantial reduction in binding, and are designated the hot spot. The hot spots on GH and GHR make contact with each other in the complex. Reprinted from [1] with permission from AAAS.

further filled by movements of a side chain on the GHR. Side chain movements resulted in breaking an H-bond and formation of a new charged H-bond near this site. There were also large movements distant from the W104 cavity. The entire interface rotated 5°, resulting in movements up to 3 Å near the periphery. This movement was accompanied by loss of four H-bonds and formation of three new H-bonds. This paper is a nice introduction to the kind of changes that might occur in the evolution of a protein–protein interface.

Although not contributing substantial binding affinity, the peripheral amino acids were suggested to serve at least two important functions. First, they would contribute to the specificity of the binding, since any protein that might bind the hot spot would also have to sterically match the non-hot spot periphery, including the peripheral region. Second, as documented especially by Cunningham and Wells [2], charged amino acids in the periphery can enhance the on rate kinetics by electrostatic steering.

6.4 Trying to predict hot spot amino acids, and protein–protein interfaces

The hot spot is not unique to the GH–GHR complex, but is a general feature of many protein–protein complexes. Clackson and Wells noted that it had already been observed in antigen–antibody complexes, where alanine scanning had shown that only 3–10 amino acids could account for most of the binding energy. Later studies found hot spots in many more protein complexes.

There has been a major research effort since the 1990s to computationally predict protein–protein interactions. This has turned out to be as difficult as predicting protein folding. Of course the two problems are related, and algorithms for both depend on force fields to measure the interactions between amino acids that make contact. In 2002 two groups developed algorithms to predict the effect of Ala mutations on protein–protein interfaces [5, 6]. Both groups trained their algorithms to recognize the $\Delta\Delta G$ of single Ala mutations on the folding of monomeric proteins. They then applied the algorithms to a set of protein complexes for which crystal structures and alanine scanning of the interfaces were available. The good news is that the predicted $\Delta\Delta G$ showed a nice correlation with the measured $\Delta\Delta G$. Guiros et al obtained an $R = 0.8$ and standard deviation of 0.66 kcal mol^{-1} [5]. Kortemme and Baker reported 80% and 84% correct identification of hot spot and neutral amino acids, where the hot spot cutoff was $\Delta\Delta G = 1$ kcal mol^{-1} [6]. The bad news is that no simple rules appeared where one could predict hot spot amino acids by simply looking at the x-ray structure. You had to run the full algorithm, and also realize that it could have a significant chance of mis-identifying the contribution to ΔG of any single amino acid.

A 2007 review covers a range of issues related to the structure of protein–protein interfaces and hot spots [7]. A more recent review focuses on antibodies binding to protein ligands, and how the binding affinity can be modulated by mutating only a small number of amino acids in the complement determining loops [8]. An ambitious project was initiated in 2013 where computational teams were challenged to predict the effect of a broad range of single amino acid mutations on the binding affinity of two target pairs. The actual binding changes had been determined experimentally but kept secret until the teams had submitted their predictions. There was much to cheer in the successes, but also much room for improvement. The results are presented in [9].

A more general community-wide challenge has been going on since 2001: Critical Assessment of Predicted Interactions (CAPRI). The goal has been to predict the binding interface of a protein pair, given the x-ray structures of the two monomers and the knowledge that they form a complex. In this semi-annual challenge the crystal structure of the challenge complex has been determined, but it is kept secret until the predictions are in. A panel of scientists evaluates how well the submissions succeeded. CAPRI prediction rounds from 2013 to 2016 have been published [10]. 'Models of acceptable quality or better were obtained for 14 of the 20 targets, including medium quality models for 13 targets and high quality models for 8 targets, indicating tangible progress of present-day computational methods in modeling protein complexes with increased accuracy.' This article and related articles from that competition will give the interested reader an entrée into this highly technical field.

The hot spot discovery requires some major revisions of the simple Chothia–Janin model. These are summarized in box 6.2.

Box 6.2. The protein-protein bond revised from Chothia-Janin to hot spots.

H-bond complementarity: The original C-J theory required that H-bonds be completed across the interface. The hot spot shows that, especially in the peripheral region, the H-bond network is plastic, and can be rearranged to compensate for a missing donor or acceptor. The H-bond plasticity is facilitated by water molecules bound in crevices in the peripheral region.

Steric complementarity: This is still very important. However, in the peripheral region the packing is less snug, and there are numerous cavities containing structured water molecules.

Buried WASA: $1,200 - 2,000$ Å^2 buried WASA is still a good indication of a reasonably tight bond. However, at 0.025 kcal mol^{-1} the hydrophobic bond energy is 2–3 times larger than the actual value ΔG_{bond}. This is consistent with the hydrophobic bonding being focused in the hot spot, which is only ~1/3 of the total interface.

References

[1] Clackson T and Wells J A 1995 A hot spot of binding energy in a hormone–receptor interface *Science* **267** 383–6

[2] Cunningham B C and Wells J A 1993 Comparison of a structural and a functional epitope *J. Mol. Biol.* **234** 554–63

[3] Amit A G, Mariuzza R A, Phillips S E and Poljak R J 1986 Three-dimensional structure of an antigen–antibody complex at 2.8 A resolution *Science* **233** 747–53

[4] Atwell S, Ultsch M, De Vos A M and Wells J A 1997 Structural plasticity in a remodeled protein–protein interface *Science* **278** 1126–8

[5] Guerois R, Nielsen J E and Serrano L 2002 Predicting changes in the stability of proteins and protein complexes: a study of more than 1000 mutations *J. Mol. Biol.* **320** 369–87

[6] Kortemme T and Baker D 2002 A simple physical model for binding energy hot spots in protein–protein complexes *Proc. Natl. Acad. Sci. USA* **99** 14116–21

[7] Reichmann D, Rahat O, Cohen M, Neuvirth H and Schreiber G 2007 The molecular architecture of protein–protein binding sites *Curr. Opin. Struct. Biol.* **17** 67–76

[8] Akiba H and Tsumoto K 2015 Thermodynamics of antibody–antigen interaction revealed by mutation analysis of antibody variable regions *J. Biochem.* **158** 1–13

[9] Moretti R *et al* 2013 Community-wide evaluation of methods for predicting the effect of mutations on protein–protein interactions *Proteins* **81** 1980–7

[10] Lensink M F, Velankar S and Wodak S J 2017 Modeling protein–protein and protein–peptide complexes: CAPRI 6th edition *Proteins* **85** 359–77

IOP Publishing

Principles of Protein-Protein Association

Harold P Erickson

Chapter 7

Cooperativity in protein–protein association and efficiency of bonds

The principles discussed here are based on Erickson 1989 [1]. That paper addressed the question of cooperative assembly, where a subunit associating with a polymer formed two bonds at once. The question was, if we know the affinity of each bond separately, what is the affinity for forming both at once? It turns out that the affinity is enormously enhanced relative to the single bonds. The key to the analysis is to correctly account for the role of intrinsic subunit entropy. We will repeat the derivation of ΔG_{bond} and ΔG_S from chapter 2, and then use it to determine the affinity of two bonds at once. This chapter will conclude with a discussion of efficiency of protein–protein bonds, and a novel analysis of cooperativity in assembly of the GHR:GH:GHR complex.

7.1 Intrinsic bond energy and subunit entropy

To understand cooperativity in protein–protein association, we need to work with free energy, not the equilibrium association constant. They are related by these equations.

$$\Delta G_A = -RT \ln(K_A); \quad K_A = e^{-\Delta G_A/RT} \tag{7.1}$$

The free energy, designated ΔG_A, describes all of the chemical and energetic factors involved in the association reaction. Note again that a favorable association means a negative ΔG_A.

The advantage of free energy is that it is additive. To utilize the additivity it is important to separate ΔG_A into two opposing energies, one favoring association and one opposing it. The two terms are the *intrinsic bond energy* and the *intrinsic subunit entropy*.

$$\Delta G_A = \Delta G_{bond} + \Delta G_S; \quad \Delta G_{bond} = \Delta G_A - \Delta G_S \tag{7.2}$$

doi:10.1088/2053-2563/ab19bach7

The term ΔG_{bond} is the intrinsic bond energy. It includes all the chemical forces operating across the protein–protein interface, which generate an overall attraction. ΔG_{bond} is a negative number because it favors the association; the stronger the bond, the larger its absolute value. The Chothia and Janin analysis discussed in chapter 3 and the hot spot in chapter 6 were directed at estimating this intrinsic bond energy and determining how specific amino acid contacts contribute to it.

The term ΔG_S is the intrinsic subunit entropy, expressed here in units of free energy. Think of this as the free energy required to immobilize a subunit in a dimer or polymer, independent of the strength or number of bonds formed. Free energy is required because entropy is lost when the subunit is immobilized. Before dimer formation each subunit has three degrees of translational and three degrees of rotational freedom. In the dimer one subunit still retains its six degrees of freedom, but the other subunit has lost its independent translation and rotation. The question is, how much free energy does it take to immobilize a subunit, to compensate for the loss of three translational and three rotational degrees of freedom? The best current value is $\Delta G_s = +6$ kcal mol^{-1}.

Chothia and Janin estimated ΔG_s to be 20–30 kcal mol^{-1}, based on a quantum mechanical calculation [2]. Erickson noted that proteins in a dimer would retain motions corresponding to ± 1 Å, and calculated a $G_s = 11$ kcal mol^{-1} [1]. However, he noted that this value was too high to fit a reasonable model of actin nucleation, and suggested 7 kcal mol^{-1} as a reasonable upper limit. Later Horton and Lewis plotted ΔG values for a range of protein complexes and the intercept gave a value of $\Delta G_s = 6$ kcal mol^{-1} [3], which we will use here. An important feature of ΔG_s is that it depends very little on the size or shape of the protein subunit. That is because these features enter the calculation as the logarithm. Therefore, the $\Delta G_s = 6$ kcal mol^{-1} can be considered universal for immobilizing a protein subunit in a dimer, oligomer or polymer.

It is useful to think about the intrinsic subunit entropy as an entropy tax. The tax is a flat rate, not progressive. The entropy tax must be paid once for any association, and the tax is the same regardless of the size of the subunit and the strength of the bond. Another way to think about the equation (7.2) is that the intrinsic bond energy must be sufficient to achieve the observed K_A, and it must also pay the entropy tax of 6 kcal mol^{-1}.

$$\Delta G_{\text{bond}} = \Delta G_A - 6 \text{ kcal mol}^{-1} \qquad (7.3)$$

Let's put this in perspective and illustrate the calculations with some numbers. A typical modest protein–protein association, such as actin assembly or a weak antibody, will have a $K_D = 10^{-6}$ M ($K_A = 10^6$ M^{-1}). In equation (7.1), R has the value 2 cal deg^{-1}·mol, and T is the absolute temperature = 300 K (= 27 °C, chosen here to make RT a round 600 cal mol^{-1} = 0.6 kcal mol^{-1}). Consider the case of actin, where $K_A \sim 10^6$ M^{-1}.

$$\Delta G_A = -RT \ln(K_A) = 0.6 \ln(10^6) = -8.3 \text{ kcal mol}^{-1} \qquad (7.4)$$

$$\Delta G_{bond} = \Delta G_A - 6 \text{ kcal mol}^{-1} = -8.3 - 6 = -14.3 \text{ kcal mol}^{-1}. \qquad (7.5)$$

The net free energy of association, -8.3 kcal mol^{-1}, is very similar in magnitude to the intrinsic entropy term, 6 kcal mol^{-1}, but of opposite sign. The intrinsic bond energy must be larger than ΔG_A because it must compensate for the entropic energy loss.

Now let's consider a higher affinity interaction, $K_A = 10^9$ M^{-1} (this would be typical for a growth factor binding its receptor).

$$\Delta G_A = -RT \ln(K_A) = -0.6 \ln(10^9) = -12.4 \text{ kcal mol}^{-1} \qquad (7.6)$$

$$\Delta G_{bond} = \Delta G_A - 6 \text{ kcal mol}^{-1} = -12.4 - 6 = -18.4 \text{ kcal mol}^{-1}. \qquad (7.7)$$

It is interesting to consider that the 1000-fold difference in K_A is achieved by only a 29% increase in intrinsic bond energy. That is because the bond energy enters the K_A as an exponential.

The primary utility of ΔG_{bond} is that components contributing to ΔG_{bond} are simply additive. This becomes especially important in the calculation of cooperativity, considered next.

7.2 Additivity of bond energies and cooperative association

What happens to K_A if you make the bond area twice as big, i.e. you double the intrinsic bond energy? A hypothetical example would be to compare binding a bivalent Ab to binding of the monovalent Fab. Specifically, let's assume we can construct an HEL dimer, with the two HEL subunits held together so they can each bind the Fab of the IgG (figure 7.1).

To make the calculation easy we introduce two assumptions, neither of which are really valid.
(a) Assume that the Ab is rigid. The Fab fragments on an IgG have considerable rotational flexibility, but for this exercise consider them rigid. Assume that the crosslinked HEL dimer is also rigid.
(b) Assume that the crosslinking presents the HEL epitopes so that the two Fabs of the rigid IgG can bind both of them without strain or distortion.

Essentially the problem now asks what happens to K_A if we double the intrinsic bond energy. Is K_A also doubled? Do you square it? Both of these quick guesses have been proposed in the literature, but they are wrong. To do the calculation correctly we need to consider the intrinsic bond energy. Most important, we have to pay close attention to the intrinsic subunit entropy.
(a) Calculate $\Delta G_{bond}^{Fab} = -RT \ln K_A - \Delta G_s = -9.7 - 6 = -15.7$ kcal mol^{-1}.
(b) Calculate ΔG_{bond}^{IgG}. This is simply $2 \times \Delta G_{bond}^{Fab} = -31.4$ kcal mol^{-1}.
(c) Calculate $\Delta G_A^{IgG} = \Delta G_{bond}^{IgG} + 6 = -31.4 + 6 = -25.4$ kcal mol^{-1}.
(d) $K_A^{IgG} = \exp(\Delta G_A^{IgG}/RT) = \exp(25400/600) = 2.4 \times 10^{18}$ M^{-1}.

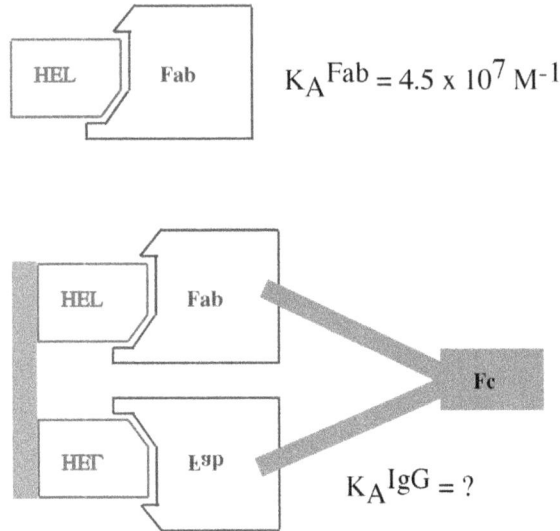

Figure 7.1. If the monovalent Fab binds monovalent HEL with $K_A^{Fab} = 4.5 \times 10^7$ M^{-1}, what is K_A^{IgG} for binding bivalent HEL to bivalent IgG?

Note that in (a) we have subtracted the ΔG_S once to obtain ΔG_{bond}^{Fab}. In (c) we have added ΔG_S once to obtain ΔG_A^{IgG}. If we had added ΔG_S twice, this would be equivalent to squaring ΔG_A^{Fab}, which is wrong.

This is an enormous enhancement from doubling the intrinsic bond energy. The binding affinity K_A goes from 4.5×10^7 to 10^{18} M^{-1}. The key to this effect, and to this calculation, is to realize that *the entropy tax is paid only once, while the intrinsic bond energy is fully counted twice.* Since the IgG is assumed to be rigid it is fully immobilized when the first Fab binds. The second Fab then binds 'for free' (no entropy tax) and all of its energy goes to increasing K_A.

Of course this is an oversimplification, since it depends critically on the assumptions of (a) rigidity of the Ab, and (b) the perfect fit to the HEL dimer. In a real Ab the second Fab will still have substantial rotational entropy after the first one has bound. This will be smaller than 6 kcal mol^{-1}, but it would have to be compensated when the second Fab binds. Also an IgG binding to two coat proteins on a virus will have to strain a bit or a lot to make the bivalent attachment. Experimental measurements of Abs binding to viruses show ~1000 × greater affinity for IgG compared to Fab [4]. This is much less than the maximum enhancement we calculated, but still a big increase.

This treatment of cooperativity was originally developed to explain the nucleation and cooperative assembly of actin [1]. Actin is a two stranded helix, and subunits make longitudinal contacts along the helix and diagonal contacts across the helix. Estimates were made for ΔG_{bond} of a subunit forming a single longitudinal or diagonal bond, and for forming both at once as it associates onto the end of a filament. The numbers provided an explanation for the very unfavorable nucleation step, where subunits made only a single bond. The reader is referred to the original article for this example [1]. Next we will apply this cooperativity analysis to the case of the trimeric complex of GHR–GH–GHR.

7.3 Analysis of cooperativity in GH–GHR association, and comments on the 'efficiency' of hydrophobic bonding

The following analysis is a simple and novel extension of the cooperativity analysis. Table 7.1 summarizes data on interface area and actual bond energy for a number of proteins, with a particular interest in the question—how does the actual bond energy compare to the Chothia–Janin prediction based on 0.025 kcal mol^{-1} $Å^{-2}$? It is comforting to see that actual bond energies are always less than the maximum hydrophobic bond energy predicted from interface area. This suggests that some imperfections in the interface may decrease the bond energy from its full hydrophobic potential. The full potential is WASA $\times$ 0.025 kcal mol^{-1} $Å^{-2}$. We can calculate a 'bond efficiency' as the ratio of the actual bond energy to the maximum hydrophobic energy available from the interface area. Table 7.1 presents this calculation for a number of protein pairs. We see that the first four proteins vary in efficiency from 70% to 46%.

The GHR–GH–GHR complex [5] presents an interesting case. There are three interfaces, sites 1, 2 and 3 (see chapter 6). The formation of the single GH–GHR dimer through site 1 is well characterized from a mutant GH that can't bond the second GHR [6]. The area of this interface, 2460 $Å^2$ (this is the area from both sides of the interface; de Vos *et al* presented the area from one side, which I have doubled) is relatively large, giving this interface an exceptionally low efficiency, 30%. The addition of the second GHR to the already-formed GH–GHR is also well characterized. This involves two interfaces, the site 2 interface on GH binding the second GHR, plus a site 3 GHR–GHR interface at the bottom of the C-terminal FN-III domains. The areas are 1800 and 1000 $Å^2$ for sites 2 and 3, for a total of 2800 $Å^2$. The 2800 $Å^2$ combined WASA of sites 2 and 3 is slightly larger than the 2460 $Å^2$, consistent with the slightly larger K_A: 5×10^9 M^{-1} for site 2 plus 3, versus 0.9×10^9 M^{-1} for 1. The bond efficiency for sites 2 plus 3 is 31%, essentially the same as the efficiency of forming site 1. We will therefore assume that each of the individual interfaces, site 1, site 2 and site 3, are 30% efficient.

Note that sites 2 and 3 form simultaneously when the second GHR binds, and this binding is high affinity. We can now use the reverse of the cooperativity argument to estimate the affinity of sites 2 and 3 separately. The calculations are presented in table 7.1, with the assumption that each of these interfaces is 30% efficient, and using the WASA determined by de Vos *et al* [5]. We see that site 2 alone is a weak bond, $K_A = 3 \times 10^5$ M^{-1} or $K_D = 3$ μM. The affinity of site 3 is so weak it is meaningless for proteins $K_A = 12$ M^{-1} or $K_D = 0.08$ M. Neither of these sites would generate any significant association alone. Yet once the GH–GHR has formed through site 1, the cooperative binding of a second GHR to a weak site 2 plus an even weaker site 3 makes a high affinity bond.

Table 7.1. Examples of protein–protein bonds. Calculation of water-accessible surface area (WASA) buried in interface, and relation to actual bond energy.

Protein–protein pair	(a) WASA Å^2	(b) Max available hydrophobic bond energy kcal mol^{-1}	(c) K_A experimental M^{-1}	(d) $\Delta G_{bond} = -RT \ln K_A$ −6 kcal mol^{-1}	(e) 'Bond efficiency' d/b
Trypsin–PTI	640 + 750 = 1390	−35	10^{13}	−18 − 6 = −24	24/35 = 69%
Barnase–barstar	803 + 707 = 1590	−40	10^{14}	−19 − 6 = −25	25/40 = 63%
Insulin dimer	530 + 600 = 1130	−28	10^{5}	−7 − 6 = −13	13/28 = 50%
HEL–mAbD1.3 (Fab)	748 + 690 = 1438	−36	4.5×10^7	−10.6 − 7 = −17.6	17.6/36 = 46%
GH–GHR1	1230 (×2) = 2460	−61.5	0.9×10^9	−12.4 − 6 = −18.4	19.7/61.5 = 30%
(GH–GHR1) |— GHR2	900 (×2) + 500 (×2) = 2800	−70	5×10^9	−13.4 − 6 = −21.4	21.4/70 = 31%
GH–GHR2	900 (×2) = 1800	−45	(3×10^5) back calculated	(−13.5) back calculated	30% assumed
GHR1–GHR2	500 (×2) = 1000	−25	(12 M^{-1}) back calculated	(−7.5) back calculated	30% assumed

Notes. (a) WASA includes both sides of the interface. (b) Maximum avail. Hydrophobic free energy is calculated as WASA times 0.025 kcal mol^{-1} Å^{-2}. (c)–(e) The values of K_A and ΔG_{bond} for GH–GHR2 and GHR1–GHR2 were back calculated from their WASA and the assumption that each had a 30% 'bond efficiency.'

7.4 Conclusions

1. Protein–protein bonds are always weaker than the maximum available hydrophobic bond energy. The actual bonds utilize from 70% to 30% of the maximum hydrophobic bond energy calculated from the WASA multiplied by 0.025 kcal mol^{-1} Å^{-2} WASA.
2. The GH:GHR bonds appear to be especially inefficient, because the WASA is much higher than for other proteins of comparable K_A. Cunningham and Wells demonstrated that the 'hot spot,' a small cluster of residues in the center of the interface contributes most of the binding energy. This hot spot cluster apparently contributes at near 100% efficiency, while the peripheral residues contribute near 0%.
3. Cooperativity applied to the GHR:GH:GHR complex explains how the small patch of site 3 interface can boost the affinity for adding the second GHR from weak (3 μM K_D for site 2 alone) to very strong (0.2 nM for site 2 plus site 3).

References

[1] Erickson H P 1989 Cooperativity in protein–protein association: the structure and stability of the actin filament *J. Mol. Biol.* **206** 465–74
[2] Chothia C and Janin J 1975 Principles of protein–protein recognition *Nature* **256** 705–8
[3] Horton N and Lewis M 1992 Calculation of the free energy of association for protein complexes *Prot. Sci.* **1** 169–81
[4] Crothers D M and Metzger H 1972 The influence of polyvalency on the binding properties of antibodies *Immunochemistry* **9** 341–57
[5] de Vos A M, Ultsch M and Kossiakoff A A 1992 Human growth hormone and extracellular domain of its receptor: crystal structure of the complex *Science* **255** 307–12
[6] Clackson T, Ultsch M H, Wells J A and De Vos A M 1998 Structural and functional analysis of the 1:1 growth hormone:receptor complex reveals the molecular basis for receptor affinity *J. Mol. Biol.* **277** 1111–18

Chapter 8

Kinetics of protein–protein association and dissociation

This chapter will address two related questions of kinetics. If a receptor is empty, how long will it take to bind its growth-factor/ligand? Once a complex is formed, how long will it take to dissociate? These questions of kinetics are important for understanding a wide range of biochemical mechanisms, and also for understanding limitations to *in vitro* purification and assays.

We have already discussed the equilibrium association/dissociation constant. We now want to break down the K_D into its kinetic constants, the on rate and off rate (figure 8.1).

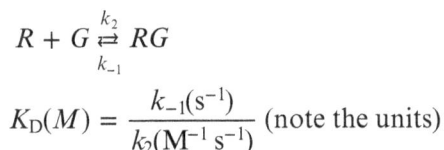

$$R + G \underset{k_{-1}}{\overset{k_2}{\rightleftarrows}} RG$$

$$K_D(M) = \frac{k_{-1}(s^{-1})}{k_2(M^{-1}\,s^{-1})} \text{ (note the units)}$$

Let's approach kinetics from the point of a receptor (R), binding its ligand (growth factor, G). We assume that the receptors on the cell surface are present in very low concentration relative to the ligand G. Kinetics can be expressed as two questions.

8.1 What is the half time of the empty receptor?

If the receptor is empty, how long will it take to be occupied? This time is inversely proportional to the second order rate constant k_2, multiplied by the concentration of G. The half time of the empty receptor is given by

$$t_{E1/2} = \ln 2 \frac{1}{k_2[G]}$$

doi:10.1088/2053-2563/ab19bach8

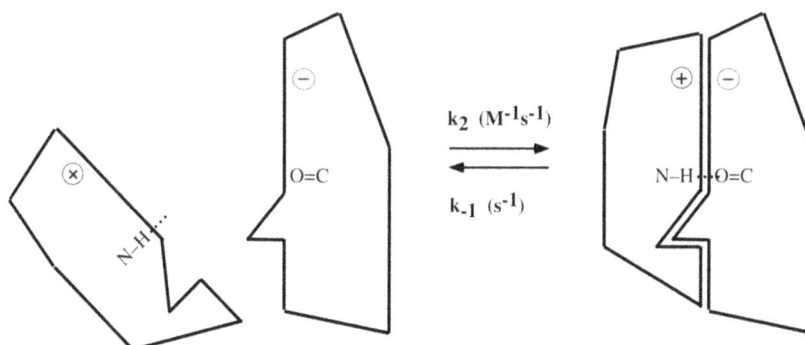

Figure 8.1. Schematic of two protein subunits associating and dissociating.

Note that $t_{E1/2}$ is not the half time for a particular receptor to be occupied. This is a stochastic process that can only be expressed as a probability. $t_{E1/2}$ is an ensemble parameter, specifically the time for half of the empty receptors in an ensemble to become occupied. The factor $\ln 2 = 0.7$ comes from the exponential in the solution of the differential equation. Importantly, the half time of the empty receptor is inversely proportional to the concentration of G. If $[G]$ is doubled, receptors will be occupied twice as fast.

8.2 What is the half time of the complex?

Once a receptor has bound a G, how long will the complex last before the G dissociates? This is again a stochastic process, and the time is inversely proportional the off rate, k_{-1}. The bound G vibrates back and forth and eventually dissociates, and the time required depends on the bond energy and vibration. The Gs in solution have no effect on this process, so their concentration does not matter. The half time of the complex is given by.

$$t_{C1/2} = \ln 2 \frac{1}{k_{-1}}$$

Importantly, the half time of the complex does not depend on the concentration of G in solution. This may be seen by looking at the figure above. The bound G fully occupies the complimentary site on R, giving no access to G in solution until the bound G fully dissociates.

8.3 The diffusion-limited rate constant for protein–protein association

Koren and Hammes [1] surveyed a number of protein associations, and found that many of them had $k_2 = 0.5$–$5 \times 10^6 \ \mathrm{M}^{-1} \ \mathrm{s}^{-1}$. To some researchers considering protein–protein association this rate seemed incredibly fast. The problem goes back to a calculation by Smoluchowski [2] that the diffusion-limited rate of encounter of smooth spheres was $k_2 = 7 \times 10^9 \ \mathrm{M}^{-1} \ \mathrm{s}^{-1}$. According to this theory, if proteins behaved as smooth spheres, and if they formed a complex every time they collided—*without regard to orientation*—they would form a complex at the rate $k_2 = 7 \times 10^9$ M^{-1} s^{-1}.

This is very fast, but consider now how it should be affected by the extremely high steric specificity of the protein–protein bond. The correct bonding requires that each subunit be oriented rotationally to about 1/1000 of its possible rotations (this would correspond to a ± 1–2 Å rotation of a surface atom). A simple argument would suggest that the kinetics of association should be slowed from the Smoluchowski limit by a factor of 1000 for each subunit. This would give a maximum, diffusion-limited rate of only $(7 \times 10^9)(1/1000)(1/1000) = 7 \times 10^3$ M^{-1} s^{-1}. Actual proteins associate 100–1000 times faster. How?

Northrup and Erickson [3] resolved this question by using Brownian dynamics, a computer simulation that treats the protein subunits as Brownian particles. It turns out the very slow rate, 7×10^3 M^{-1} s^{-1}, would be appropriate if proteins were in a vacuum. In a vacuum, if they bumped in to each other in the wrong orientation they would bounce apart and never see each other again. The Smoluchowski limit, modified by the (1/1000)(1/1000) geometry would apply. However, for proteins in water, if they collide in the wrong orientation, they simply diffuse a short distance apart. Because of this 'diffusive entrapment' the proteins have a high probability of rotating to a new position and bumping into each other again and again. The Brownian dynamics simulation estimated that $k_2 = 2 \times 10^6$ M^{-1} s^{-1} would be the generic, diffusion limited rate constant for protein–protein association in water.

It is important to note that some protein associations occur 10–100 times slower. These slow complexes have an additional energy barrier to complex formation. Also, some associations are also 10–100 times faster. Most of these very fast reactions have been characterized as due to electrostatic steering. Charge groups on the surface of the subunits steer them into correct alignment as they are approaching each other. This is discussed later.

The value $k_2 = 2 \times 10^6$ M^{-1} s^{-1} also applies to many small molecule ligands binding to a protein. It is a good number to remember.

8.4 Half time of the empty receptor and the complex—guessing the kinetics

Often we don't have experimental data for the kinetic constants, but we do know the equilibrium dissociation constant K_D. In this case we can make a guess of the kinetics by assuming that the on rate is the diffusion limited $k_2 = 2 \times 10^6$ M^{-1} s^{-1}. Consider first the half time of the empty receptor. As stated above this depends on the concentration of G and on k_2.

$$t_{E1/2} = 0.7 \frac{1}{k_2[G]}$$

The second order association rate constant, k_2, is a key parameter. If we don't know k_2 we can start by assuming the association is diffusion limited, so set $k_2 = 2 \times 10^6$ M^{-1} s^{-1}.

$$t_{E1/2} = 0.7 \frac{1}{(2 \times 10^6)[G]}$$

If the protein–protein association is diffusion limited, the half time of the empty receptor depends only on the concentration of free G. It does not depend on k_{-1} or K_D. The box tabulates some values. If the concentration of G is 1 nM, an empty receptor will be occupied in 350 s. If $[G]$ is increased to 10 nM, the receptor will be occupied ten times faster. If one considers a much more abundant protein like actin, whose concentration is ~10 μM, subunits will add to an actin filament with a half time of 35 ms.

	$[G]$	$t_{E1/2}$
GrFac	10^{-9} M	350 s
	10^{-8} M	35 s
(Actin)	10^{-5} M	0.035 s

Consider next the half time of a complex. As stated above this is given by $t_{C1/2} = 0.7 (1/k_{-1})$, and it would seem that we would need to know this rate constant. But here again we can make a guess, assuming that the association rate may be diffusion limited, where $k_2 = 2 \times 10^6$ M^{-1} s^{-1}. If we know K_D and can guess k_2, we can estimate k_{-1} and $t_{C1/2}$.

$$t_{C1/2} = 0.7 \frac{1}{k_{-1}} = 0.7 \frac{1}{K_D k_2} = 0.7 \frac{1}{K_D(2 \times 10^6)}$$

The table gives some numerical examples, ranging from the very strong complex of trypsin–trypsin inhibitor, down to the moderate affinity of an actin monomer dissociating from the end of an actin filament.

	K_D	$t_{C1/2}$
Trp–TrpInhib	10^{-13}	3.5×10^6 s (40 days)
GH–GHR	10^{-9}	350 s (6 min)
actin	10^{-6}	0.35 s

It is interesting now to consider the association–dissociation as a cyclic event. Consider a growth factor binding to its receptor. If the concentration $[G]$ is 10^{-9} M, the half time of the empty receptor is 350 s. Once a complex is formed its half time, determined by $K_D = 10^9$ M, is also 350 s. This is another way of saying that when $[G]$ is equal to K_D, the receptor is 50% occupied. If the concentration of G is now increased ten-fold to 10^{-8} M, the half time of the complex remains unchanged at 350 s, but the empty receptor now binds ligand in only 35 s. The receptor is on average ~90% occupied.

8.5 Proteins can associate much slower and much faster than the diffusion-limited rate

We should not overemphasize the generic diffusion-related on rate, since actual k_2 rate constants can span a wide range. Pang and Zhou [4] examined four cytokine–receptor pairs. Each of the cytokines and receptors had the same fold as hGH and hGHR, and the complexes superimposed with RMSD $\sim$4 Å (except for IL4, which bound in a different orientation). In spite of the similarity in structure, the on-rate constants for association span a range of $\sim 10^4$. From the structures of the cytokine–receptor complexes, and applying their own transient-complex theory, Pang and Zhou concluded that 'the vast differences in receptor-binding rate constants of the four cytokines arise mostly from the differences in charge complementarity.' This was also the conclusion of Cunningham and Wells [5], who observed that mutating charged amino acids peripheral to the hot spot affected the on rate.

EPO (erythropoietin)	4.0×10^8 M^{-1} s^{-1}
IL4 (interleukin-4)	1.3×10^7
hGH (growth hormone)	3.2×10^5
PRL (prolactin)	8.0×10^4

The reader is referred to Pollard and De La Cruz for a brief treatment of kinetics that reinforces many of the points made in this chapter and extends into practical issues of measurement and interpretation [6].

References

[1] Koren R and Hammes G G 1976 A kinetic study of protein–protein interactions *Biochemistry* **15** 1165–71
[2] Smoluchowski M v 1916 Drei Vorträge über diffusion, Brownsche Molekularbewegung und Koagulation von Kolloidteilchen *Physik. Zeitschr.* **17** 585–99
[3] Northrup S H and Erickson H P 1992 Kinetics of protein–protein association explained by Brownian dynamics computer simulation *Proc. Natl. Acad. Sci. USA* **89** 3338–42
[4] Pang X, Qin S and Zhou H X 2011 Rationalizing 5000-fold differences in receptor-binding rate constants of four cytokines *Biophys. J.* **101** 1175–83
[5] Cunningham B C and Wells J A 1993 Comparison of a structural and a functional epitope *J. Mol. Biol.* **234** 554–63
[6] Pollard T D and De La Cruz E M 2013 Take advantage of time in your experiments: a guide to simple, informative kinetics assays *Mol. Biol. Cell* **24** 1103–10

IOP Publishing

Principles of Protein–Protein Association

Harold P Erickson

Chapter 9

Techniques for measuring protein–protein association—use and misuse of ELISA

Directed reading:

Tangemann K and Engel J 1995 Demonstration of non-linear detection in ELISA resulting in up to 1000-fold too high affinities of fibrinogen binding to integrin αIIbβ3 *FEBS Lett.* **358** 179–81 [1].

This chapter will discuss several techniques for measuring protein–protein association in the laboratory, with the goal of obtaining a measure of K_D. I will first discuss methods that can be used *in vivo* for a qualitative screen of protein association, and then discuss *in vitro* assays that can be used to confirm association and quantitatively determine K_D. I will then discuss in detail one of the simplest and most popular techniques, the ELISA (Enzyme Linked ImmunoSorbant Assay). The ELISA is a great technique for qualitatively demonstrating a binding reaction, but it is often misused to interpret a quantitative K_D. The paper of Tangemann ane Engel [1] clearly demonstrates the inadequacy of the simple ELISA. I will conclude by discussing a competitive ELISA, which can give a true measure of the K_D in solution by incorporating extra steps and calibration [2].

9.1 Qualitative assays to screen for protein–protein association *in vivo*

Several techniques have been developed to screen for binding partners *in vivo*, without attempting to measure K_D.

- Yeast two hybrid (Y2H) [3] fuses two potentially associating proteins to separate domains of GAL4. If the proteins associate they bring the GAL4 domains together, inducing transcription of beta-galactosidase and producing a blue color. A large library of cDNAs can be used for one of the pair, screening for proteins that bind the other one.

- Tandem Affinity Purification (TAP) [4] uses a pair of affinity tags to purify one specific protein by two successive binding columns. Mass spectrometry (MS) is then used to assay for proteins bound to it. TAP requires that the complex remain together during the two-step purification so it will mainly identify very strongly associating proteins. TAP has been applied to genomic screens by tagging a broad array of proteins.
- BioID [5] fuses a promiscuous biotin ligase to a protein of interest (bait), and this generates activated biotin that covalently labels proteins within 10–50 nm, in particular those in complex with the bait. The labeled proteins are purified by binding to streptavidin and identified by MS. The labeling takes place over several hours, so this technique is most useful for stable cellular processes. Substantial improvements in speed and extension to RNA and DNA have been reviewed [6].
- APEX2 [7] uses a genetically engineered peroxidase to achieve a similar biotinylation of proteins bound to a bait protein of interest. It promises a higher resolution range than BioID and a labeling time of minutes instead of hours.

9.2 Quantitative methods for measuring the K_D of protein–protein association

The techniques discussed above are used to screen for protein–protein interactions *in vivo*. They have many false negatives and false positives, so associations need to be confirmed by purifying the proteins and assaying them *in vitro*. In addition to confirming an association, the *in vitro* assays can quantitate the affinity of the interaction and determine the K_D.

Using the nomenclature of chapter 2, we assume two proteins R (receptor) and G (growth factor) associate to form the complex RG. One can determine the K_D of the association if there is an assay for the complex RG, or for free R or free G. The emphasis here is on free R or G, as opposed to the total amount:

$$[G_{\text{free}}] = [G_{\text{TOT}}] - [RG] \text{ and } [R_{\text{free}}] = [R_{\text{TOT}}] - [RG].$$

If one has an assay for $[G_{\text{free}}]$ or $[R_{\text{free}}]$ or $[RG]$, and one knows the total amount of G and R in the assay, one can calculate the others. Experimental systems for measuring protein–protein association are based on assays for one of these components. Here I describe three assays that can be performed in a general biochemistry lab without specialized equipment, and then discuss assays that need special equipment and expertise.

9.3 Assays that can be done in most laboratories

- Tryptophan fluorescence is frequently sensitive to the environment. If a Trp is in or near the binding interface the fluorescence may be either enhanced or quenched upon binding. Most proteins have several Trp, which can compromise the assay if the signal from a Trp at the interface is drowned by emission from others. However, it is always worth checking to see if Trp fluorescence is altered by association, since this then provides a simple and rapid assay, and it requires no modification of the protein. Sometimes a Trp can be introduced by mutation and it will report association. This was ideal for FtsZ, which has no natural Trp [8].

- Environment sensitive fluorophores. It is sometimes possible to label a protein with a fluorophore that changes its fluorescence in response to binding. For example, pyrene–iodoacetamide labels one reactive Cys in actin, and its fluorescence increases 20-fold upon assembly [9], providing a valuable assay to study kinetics and nucleation. In some cases a Cys can be engineered at a location near the binding site and coupled to a naphthalene fluor that reports association [10].
- Pull-down assays. The goal here is to set up a reaction in which one component, R, is dilute in solution, and the other, G, is coupled to beads at a range of concentrations spanning from below to above the K_D. The reaction is allowed to come to equilibrium and the beads are pelleted, bringing down all G and whatever fraction of R is bound at that concentration of G. The supernatant is then assayed for $[R_{free}]$. See Pollard [11] for detailed discussion and an example (his figure 1).

9.4 Assays requiring specialized equipment and expertise

The following assays are frequently available in major research institutions. The instrumentation is in the range of $100–500k, and specialized expertise is needed to run the equipment and interpret the results.

- Isothermal titration calorimetery. ITC measures association of native proteins free in solution, the ideal situation. However, it is limited to K_D in the range mid nM to low mM. Also, it requires a lot of protein, 100 µl of one protein at 200 times the K_D.
- Fluorescence depolarization. This also measures binding of components free in solution. It requires a fluorescent label on one component, so some care must be taken to assure that the label does not affect the binding.
- Surface Plasmon Resonance (SPR) and Bio-Layer Interferometry (BLI). These techniques immobilize one component on a surface and expose this to the second by flow or immersion. Binding increases the refractive index of the surface layer, which alters a laser signal. SPR and BLI are typically used to measure on–off kinetics, from which K_D can be determined. Having one component immobilized can cause deviations from solution-phase binding. Schuck and Minton [12] pointed out the errors that could be caused by limited mass transport, and how to control for them. However, these are frequently ignored.
- MicroScale Thermophoresis (MST). This technique measures the diffusion of protein molecules in a small thermal gradient, and is based on the practical observation that the diffusion of one component is usually altered by binding even a small partner. One component needs to be labeled with a fluorescent tag. The binding is measured with both components in solution, and requires

a volume of only 4 μl. See [13] for applications to various protein pairs and a comparison to SPR.

9.5 Fitting the binding data to determine K_D

Whatever the assay, you will end up with data for the concentration of the complex GR with one of the components (R_{TOT}) at a fixed concentration, and the other (G_{TOT}) varied over a range that hopefully surrounds the K_D. Recall from chapter 2 the quadratic formula

$$[GR] = \frac{([R_{TOT}] + [G_{TOT}] + K_D) - \sqrt{([R_{TOT}] + [G_{TOT}] + K_D)^2 - 4[R_{TOT}][G_{TOT}]}}{2}$$

The appendix to chapter 2 describes how to use excel solver to fit the data to obtain the best estimate of K_D.

The review by Pollard [11] is an excellent survey of 'Simple and Informative Binding Assays.' The 2015 Springer Protocols 'Protein–protein interactions: methods and applications' (*Methods in Molecular Biology* vol 1278) provides detailed discussion and protocols for many methods for assaying protein–protien interactions, as well as several chapters discussing modern conceptual issues.

9.6 ELISA—use and misuse

Credit for inventing the ELISA is generally given to Engvall and Perlman, 1972 [14]. Their assay, and most applications in the 1970s, was used to identify antibodies in serum against various antigens, and assay their affinity semi-quantitativly. In the 1980s the ELISA began to be used as a test for association of any two proteins. The conventional ELISA as used today is described in the Box. Step 5 is the key to widespread use. One can purchase HRP conjugated goat anti-rabbit antibody and use this in the final step for any protein for which you have a rabbit antibody.

Conventional ELISA to demonstrate association of protein X to protein Y.

1. Coat plastic wells (96-well format) with protein X. Adsorb non-specifically 2–18 h. Wash off unadsorbed protein.
2. Coat with BSA to block plastic not coated by protein X. Wash 3×.
3. Add protein Y, incubate 2+ hrs to let Y bind to the plastic-bound X. Wash 3×.
4. Add (primary) rabbit antibody against protein Y, incubate 2 h. Wash 3×.
5. Add (secondary) antibody detection conjugate: this is a goat antibody against rabbit IgG. The goat Ab is conjugated to horse radish peroxidase HRP. Wash 3×.
6. Add reagents to develop HRP (brown color).
7. Read (quantitate) the brown color.

The wash steps and incubation with antibodies pose a theoretical concern about the ELISA. Since the initial X–Y association, as well as the binding of primary and secondary antibodies, are reversible reactions, one should question whether the association will be reversed during the long incubations and washes. For a K_A ~4.5 × 10^7 of a typical monoclonal antibody, assuming a diffusion-limited on rate (chapter 8), the half time of the complex is 16 s. Each wash is typically ~30 s or more, so one would expect the antibodies to be largely eluted during the washes. More important, one would expect the initial X–Y association to completely reverse during the 2 h incubations with antibodies. However, this seems not to be a major problem. ELISA works!

How can the conventional ELISA avoid the reversible detachment of ligands during the antibody incubations and washes? Surprisingly, I have not found this addressed in the literature. I will offer my own speculation, based on the concept of cooperativity (chapter 7). This showed that a protein–protein bond of modest affinity could be enormously enhanced if one added even a small amount of additional bonding surface. In the case of the ELISA this extra bonding surface might be achieved by nonspecific contact with adjacent proteins on the substrate, or to small bits of plastic not blocked by BSA. This contact would be too weak to provide any significant binding of protein Y to the control BSA-coated well. But if the well is first coated with protein X, for which Y has a significant affinity, then Y will initially bind reversibly to X. Following this initial binding, Y may contact an adjacent non-specific protein or plastic, and this will form a cooperative bond of very high affinity. The same scenario would hold for the primary and secondary antibodies. They are initially bound by their specific association of moderate affinity, but this is enhanced by non-specific contacts to adjacent protein or plastic.

9.7 A simple ELISA can over- or under-estimate the K_D by orders of magnitude

The ELISA has been used to identify thousands of protein–protein associations and is excellent for a semi-quantitative analysis. False positives are rare or non-existent, and strong binding proteins generally produce color at lower concentrations than weak binding ones. It seems natural to plot the color versus concentration of the soluble protein and use this curve to estimate the K_D. A quick alternative is to take the concentration producing 50% maximum color as an indication of the K_D. This simple interpretation of ELISA to estimate K_D is found in thousands of papers published in the most rigorous biochemical journals. But it turns out that the K_D estimated from the simple ELISA can be two or three orders of magnitude different from the true K_D. *And the error can be in either direction.*

Tangemann and Engel [1] documented this error by comparing three types of ELISA to measure the association of fibrinogen to the integrin αIIbβ3. All assays used αIIbβ3 coated on plastic wells, and soluble fibrinogen as ligand. The different assays included one, two or three steps for detection of bound fibrinogen. The most direct assay used radiolabeled fibrinogen, where the binding could be detected directly following a quick wash, a one-step assay. The two-step assay used biotin-labeled fibrinogen, which was detected by incubating with HRP-conjugated strep-tavidin, followed by HRP color development. The third assay was a conventional ELISA, which used a three-step detection: bound fibrinogen was labeled with a primary rabbit IgG against fibrinogen, and this was labeled with a secondary HRP-conjugated goat anti-rabbit IgG, followed by HRP color development. The curves of color as a function of fibrinogen were fit to the binding equation (equation (Appendix) in chapter 2) to determine the apparent K_D. The results of the three assays are presented in table 9.1.

The 47 nM K_D determined by the one-step radioligand assay agreed well with a previous assay using an independent direct technique to measure bound fibrinogen. The two-step biotin–streptavidin assay gave a K_D 10 or 100 times lower, indicating a much higher affinity. Note also that the apparent affinity in this assay depended strongly on the concentration at which the integrin was coated on the plastic. Most scientists using ELISA ignore this variable and run only a single substrate coating concentration. The three-step conventional ELISA indicated a K_D that was 1000 times too low.

The concern raised above about dissociation during the lengthy incubation and washing steps would predict that the ELISA would underestimate the affinity of binding, returning an apparent K_D that was higher than the true K_D. Tangemann and Engel found the opposite, that the ELISA indicated a binding that was 1000× higher affinity than the true K_D. The physical–chemical reason for this discrepancy is not known.

Another test comparing ELISA to the true K_D was presented by Peleg-Shulman, Schreiber and colleagues [15]. They were studying the binding of interferon β (INFβ) to the ectodomain of its receptor. Mutants of INFβ had been discovered previously that enhanced its binding to the receptor (negative $\Delta\Delta G$) and these were further characterized (table 9.2).

The true K_D in solution was measured by a combination of stopped flow (for k_2) and plasmon resonance (for k_{-1}). For the ELISA the receptor, wt or mutant, was coated on the plastic, and INFβ was in solution at different concentrations. The ELISA used a conventional secondary antibody detection. The number reported is the concentration of INFβ that gave 50% maximum color, which is often taken as an estimate of the K_D. The authors noted that the 'ELISA 50%' was about 160 times above the true K_D, which means that the ELISA here is indicating a weaker interaction than the true K_D. This is in the opposite direction to the discrepancy found by Tangemann and Engel [1]. Again, no one has explored the mechanism.

This study is not all bad news for the ELISA. It actually suggests a potential for quantitative interpretation. The encouraging point is that the 'ELISA 50%' was 160–180 times the solution K_D for all mutants tested. This suggests that the ELISA may

Table 9.1. Comparison of a conventional 3-step ELISA with 2- and 1-step assays.

Apparent dissociation constant K_D (nM) determined by the different assays			
Integrin coating concentration	ELISA	Biotin-streptavidin	Radioligand
5 nM	0.09	5	47
21 nM	0.045	0.43	47

Table 9.2. Binding of INFβ to interferon receptor (wt and mutants) measured in solution (stop flow and plasmon resonance) and by ELISA (data from [15]).

Receptor mutant	K_D solution: nM	K_D relative to wt	ELISA 50%: nM	ELISA 50% relative to WT
wt	0.1	1.0	16	1.0
H78A	0.024	0.24	8	0.5
N100A	0.009	0.09	1.6	0.1
H78A/N100A	0.002	0.02	0.35	0.02

be accurate in reporting the relative change in K_D if one protein is constant and the other varies only by point mutations. Although this needs to be tested more broadly, it would suggest that ELISA may be able to determine the $\Delta\Delta G$ in an Ala scanning study.

In conclusion, the simple ELISA can be orders of magnitude off from the true K_D, and in either direction. However, if one is comparing relative binding of two proteins that are altered by simple point mutations, the ELISA may be able to report the $\Delta\Delta G$.

9.8 A competitive ELISA can be used to measure the true K_D

Frigat *et al* [2] designed a competitive ELISA and demonstrated that it could be used to measure the true K_D. In this assay protein X is coated on plastic as above, and the assay is first calibrated with dilutions of protein Y, and developed in the conventional manner. The goal of this step is to identify a range of protein Y that gives a linear response in the conventional ELISA. One then selects a concentration of protein Y near the top of this linear range, and sets up competition samples in separate tubes. These all have the same concentration of protein Y, and increasing concentrations of soluble protein X, hopefully bracketing the K_D. These competition samples are incubated in tubes separate from the ELISA wells for an hour or more to achieve equilibrium. Then the samples are added to ELISA wells and incubated to allow the free protein Y to be captured by the substrate protein X. This is then developed by primary and secondary antibodies in the conventional manner. The important point is that protein Y that is bound to protein X in the equilibrium cannot bind to the substrate protein X. The assay will therefore determine the amount of free protein Y, which will vary from 100% with no soluble protein X, to 0% when protein X is in great excess over the K_D. A plot of free Y versus total soluble X can be used to determine K_D.

A minor drawback to this competitive ELISA is that one has to calibrate the ELISA initially, to determine a concentration of protein Y that gives a linear response over a ten-fold dilution. This may require testing a few concentrations of protein X to coat the substrate. Once a linear assay is demonstrated, the competition step gives a measure of the equilibrium K_D with both proteins in solution, and no tags or modifications on either. A major advantage of the competition ELISA is that it requires no specialized equipment and associated expertise.

References

[1] Tangemann K and Engel J 1995 Demonstration of non-linear detection in ELISA resulting in up to 1000-fold too high affinities of fibrinogen binding to integrin aIIbb3 *FEBS Lett.* **358** 179–81

[2] Friguet B, Chaffotte A F, Djavadi-Ohaniance L and Goldberg M E 1985 Measurements of the true affinity constant in solution of antigen–antibody complexes by enzyme-linked immunosorbent assay *J. Immunol. Meth.* **77** 305–19

[3] Fields S and Song O 1989 A novel genetic system to detect protein–protein interactions *Nature* **340** 245–46

[4] Rigaut G, Shevchenko A, Rutz B, Wilm M, Mann M and Seraphin B 1999 A generic protein purification method for protein complex characterization and proteome exploration *Nat. Biotechnol.* **17** 1030–32

[5] Roux K J, Kim D I, Raida M and Burke B 2012 A promiscuous biotin ligase fusion protein identifies proximal and interacting proteins in mammalian cells *J. Cell. Biol.* **196** 801–10

[6] Trinkle-Mulcahy L 2019 Recent advances in proximity-based labeling methods for interactome mapping. F1000Res 8.

[7] Hung V, Udeshi N D, Lam S S, Loh K H, Cox K J, Pedram K, Carr S A and Ting A Y 2016 Spatially resolved proteomic mapping in living cells with the engineered peroxidase APEX2 *Nat. Protoc.* **11** 456–75

[8] Chen Y and Erickson H P 2011 Conformational changes of FtsZ reported by tryptophan mutants *Biochemistry* **50** 4675–84

[9] Cooper J A, Walker S B and Pollard T D 1983 Pyrene actin: documentation of the validity of a sensitive assay for actin polymerization *J. Muscle Res. Cell Motil.* **4** 253–62

[10] Siarheyeva A and Sharom F J 2009 The ABC transporter MsbA interacts with lipid A and amphipathic drugs at different sites *Biochem. J.* **419** 317–28

[11] Pollard T D 2010 A guide to simple and informative binding assays *Mol. Biol. Cell* **21** 4061–67

[12] Schuck P and Minton A P 1996 Analysis of mass transport-limited binding kinetics in evanescent wave biosensors *Anal. Biochem.* **240** 262–72

[13] Seidel S A *et al* 2013 Microscale thermophoresis quantifies biomolecular interactions under previously challenging conditions *Methods* **59** 301–15

[14] Engvall E and Perlmann P 1972 Enzyme-linked immunosorbentassay, ELISA *J. Immunol.* **109** 129–35

[15] Peleg-Shulman T, Roisman L C, Zupkovitz G and Schreiber G 2004 Optimizing the binding affinity of a carrier protein: a case study on the interaction between soluble ifnar2 and interferon beta *J. Biol. Chem.* **279** 18046–53

Chapter 10

Fibronectin, the FNIII domain, and artificial antibodies

10.1 Fibronectin, cell adhesion and RGD

Directed reading of:

Pierschbacher M D and Ruoslahti E 1984 Cell attachment activity of fibronectin can be duplicated by small synthetic fragments of the molecule *Nature* **309** 30–3 [1].

In 1984, 11 years before Clackson and Wells defined the hot spot (chapter 6), Pierschbacher and Ruoslahti found evidence that a small tri-peptide could dominate the binding of a domain of fibronectin (FN) to its integrin receptor [1]. Before discussing this paper we will cover some background on FN.

FN is a large extracellular matrix molecule that forms the primitive matrix fibrils in embryonic development and wound healing. The primitive FN matrix serves as a scaffold for the more definitive collagen matrix. In addition to assembling the matrix fibrils, FN molecules provide attachment sites for cell adhesion.

Figure 10.1 shows the FN molecule in the context of its function in cell adhesion. We will first address the FN molecule. FN was one of the first large proteins discovered to have a modular structure, i.e. a string of homologous small protein domains. Starting from the N terminus, there are six domains of type FN-I, two FN-II and three more FN-I. The middle part of the molecule comprises a string of 15–17 FN-III domains (the domains labeled A and B can be included by alternative splicing). These are followed by three more FN-I domains and a pair of Cys that couple two FN subunits into a covalent dimer.

FN-III sounds highly specialized, the third type of domain discovered in fibronectin. But FN-III is actually one of the most common protein domains, ranking with Ig and EGF domains as one of nature's favorite building blocks. The FN-III domains in FN have only a limited sequence identity, ~15%–25%. FN-III domains in other proteins have a similar limited sequence identity. However, all FN-III domains share an identical protein fold, a beta sandwich with four strands on one

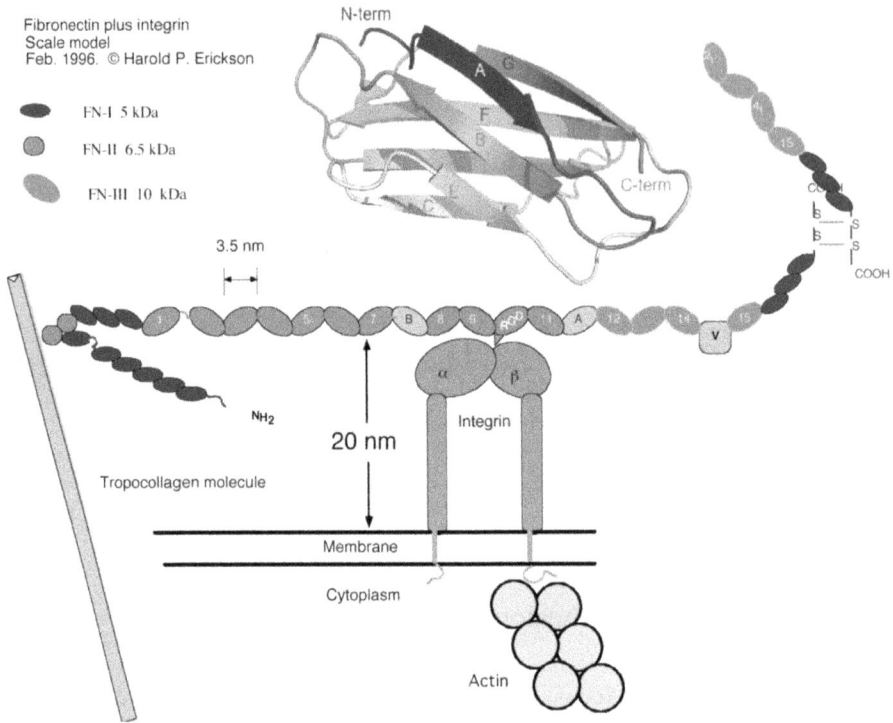

Figure 10.1. Diagram of an FN molecule and its integrin receptor. All of the molecules/domains are drawn to scale. Cyan domains are alternatively spliced. Top center shows a ribbon diagram of an FN-III.

side and three on the other. This is very similar to the structure of Ig domains, which differ in having one strand swapped to the other side; also Ig domains have internal disulfides, while FN-III domains have no disulfides.

FN-III domains have various functions. Some of them may serve as spacers, separating specific binding functions. This is probably the case for many domains of FN, tenascin and the giant muscle protein titin. Some FN-III domains have specific binding functions. The 10th FN-III domain of FN binds integrins to mediate cell adhesion. The growth hormone receptor (GHR), uses two FN-III domains to bind its ligand (chapter 5). A number of other receptors use FN-III domains for binding ligands or for linkers.

From experiments in the 1990s, it was known that FN had several binding activities. Cell adhesion experiments suggested that some part of FN bound to specific receptors on the cell surface. Binding to gelatin (denatured collagen) suggested that FN might link cells to the collagen matrix. Before the sequence of FN revealed its modular domain structure, scientists had used limited proteolysis to break FN into large fragments and map the binding activities. Gelatin binding mapped to the N-terminal fragment, now known to comprise FN-I and FN-II domains. Cell adhesion mapped to a large central chymotryptic fragment of 120 kDa, which would contain ~12 FN-III domains. Prior to the RGD discovery in the

Code	Sequence	Concentration for 50% cell attachment
Fibronectin	*a*	$0.10\ \text{nmol ml}^{-1}$
IV	Y A V T G R G D S P A S S K P I S I N Y R T E I D K P S Q M (C)	0.25
IVA	V T G R G D S P A S S K P I (C)	1.6
IVB	S I N Y R T E I D K P S Q M (C)	>50.0
IVA1	V T G R G D S P A (C)	2.5
IVA2	S P A S S K P I S (C)	>50.0
IVA1a	V T G R G D (C)	10.0
IVA1b	G R G D S (C)	3.0
IVA1c	R G D S P A (C)	6.0
RVDS	R V D S P A (C)	>50.0

Figure 10.2. These FN peptides were tested by Pierschbacher Ruoslahti for cell attachment [1]. Each peptide has a C-terminal Cys that covalently coupled the peptide to IgG protein adsorbed on the plastic substrate. The peptides were coupled at increasing concentration, and the concentration for 50% cell attachment is given in the right-hand column. If this value was >50 there was no attachment observed. The smaller the value the higher the attachment activity. Reprinted from [1] with permission from Springer Nature. Copyright 1984.

present paper, Pierschbacher and colleagues mapped the cell adhesion activity more precisely by further digesting the 120 kDa fragment with pepsin. They obtained a small ~11.5 kDa fragment that retained cell adhesion activity. This fragment, now identified as the tenth FN-III domain, was resistant to any further digestion. To study it in finer detail Pierschbacher and colleagues determined its sequence by laborious protein sequencing. Then they constructed four synthetic peptides each covering ~1/4 of the sequence. Three of these peptides had no adhesion activity, but one 30-amino acid peptide, from the C terminus, mediated cell adhesion when attached to a substrate.

The 1984 Nature paper presented here started with this 30-amino acid peptide and tried to map the activity more precisely (figure 10.2). The starting peptide, designated IV, was divided into two halves and these were tested for cell adhesion. The N-terminal half, IVA, retained substantial adhesion activity, while IVB was completely inactive. IVA was then split in two, and only IVA1 had activity. This peptide was shortened at each end, and the highest activity found was for the penta-peptide GRGDS. A single amino acid substitution R → V completely destroyed the activity. The conservative substitution D → E also completely abrogates adhesion, and RGE has become the standard negative control. Later work confirmed that the tripeptide RGD was the key active site, while the flanking amino acids could modulate the activity.

Pierschbacher and Ruoslahti used two assays to demonstrate the activity. The first assay was a positive one, showing that the small peptides containing the RGD sequence could mediate cell adhesion when they were coupled to the substrate. They determined the concentration of peptide that would provide 50% maximal adhesion when coupled to a substrate (see the paper for details of the coupling method). The direct adhesion data are tabulated in figure 10.2).

They followed this with a negative assay, showing that the peptides in solution could inhibit adhesion to native FN on the substrate. This is shown in figure 10.3.

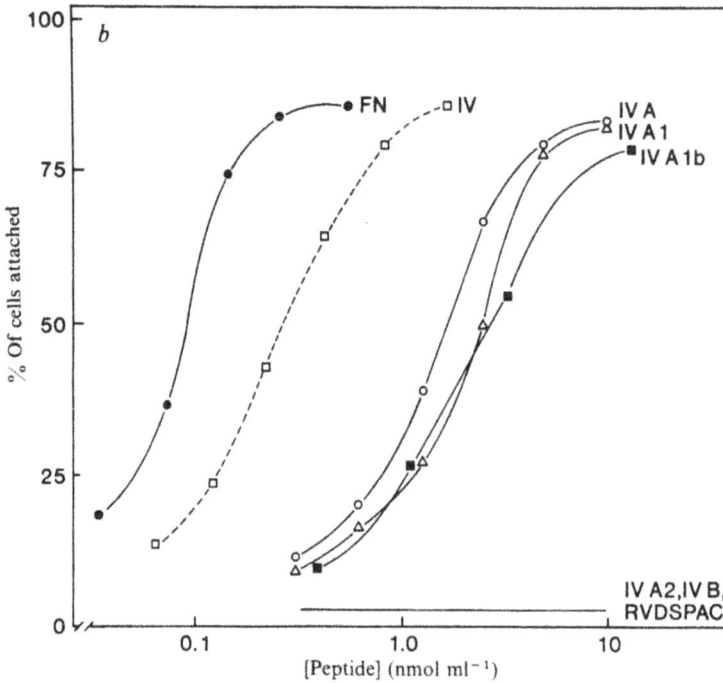

Figure 10.3. Plastic wells were coated with 5 µg ml^{-1} FN, and cell adhesion was competed by increasing concentration of soluble peptide. RGDS and GRGDSP blocked cell adhesion at concentrations ~0.3–1.0 mM, while the control peptides did not. Reprinted from [1] with permission from Springer Nature. Copyright 1984.

One might be concerned that the inhibition required a high (mM) concentration of peptide. This is because the peptide in solution is flexible and adopts multiple conformations. If only a small fraction of these conformations are active in binding the integrin receptor, that would explain the need for a high concentration. Consistent with this, later studies found that cyclic peptides containing the RGD sequence had higher activity, i.e. they were effective at lower concentrations. Cyclizing the peptide locks it into a favorable conformation.

When this study was published in 1984 it was not clear how a tripeptide could have such a large influence on the protein–protein bond. The answer came with the crystal structure of a segment of FN containing the RGD sequence [2] (figure 10.4). This segment, called FN7–10, contained FN-III domains 7–10, where domain 10 has the RGD. The RDG is on the FG loop, between the F and G beta strands. This loop in domain 10 is three amino acids longer than the FG loop in all other FN-III domains, which lets the RGD sequence protrude significantly from the surface. A crystal structure of integrin αVβ3 bound to FN10 showed the RGD loop inserted into a cleft between the alpha and beta subunits of the integrin [3]. This is reminiscent of the Gln121 of HEL inserting into a cleft between the VL and VH chains of the D1.3 antibody (reference [4] and chapter 4). H-bonding amino acids inserting into a cleft can have a powerful effect on binding affinity.

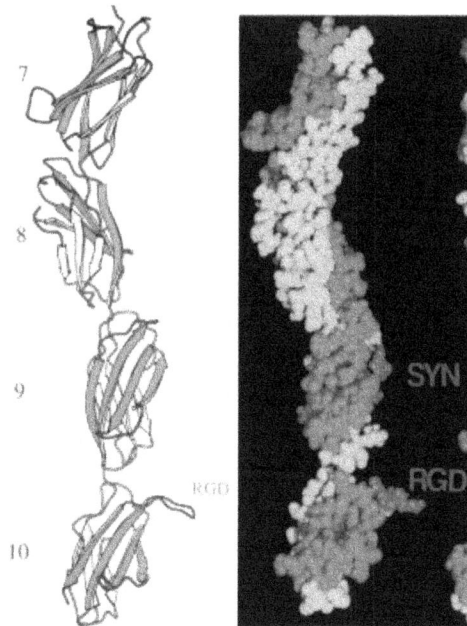

Figure 10.4. The structure of FN7–10 showing the RDG on a loop protruding from FN-III domain 10. The beta strand structure is shown on the left, and a spacefilling model is on the right. Reprinted from [2], copyright 1996, with permission from Elsevier.

It might seem that the use of peptides to identify, mimic and inhibit protein–protein bonds would be limited to special cases like the RGD inserting into the cleft. However, the paradigm of peptide mimics has been extended to many different proteins and has been remarkably successful in identifying binding sites and potential inhibitory peptides. See [5] for a description of a miniaturized high-throughput system for screening large numbers of peptides.

10.2 Antibody mimics—creating novel binding activities in a neutral protein framework

Directed reading of:

Koide A, Bailey C W, Huang X and Koide S 1998 The fibronectin type III domain as a scaffold for novel binding proteins *J. Mol. Biol.* **284** 1141–51 [7].

The immune system has the ability to generate antibodies against almost any antigen, protein or otherwise. As discussed in chapter 4, the human IgG is a complex arrangement of Ig domains. Even the monovalent Fab comprises four Ig domains, and the antigen-binding site includes CDRs from the VL and VH domains. However, four or even two domains are not really needed for high affinity binding. The VH domain alone can have most of the binding activity and also generate a broad range of antigen specificity. Camels confirm this point by having a class of antibodies with only a heavy chain. The VH domain of these camel antibodies,

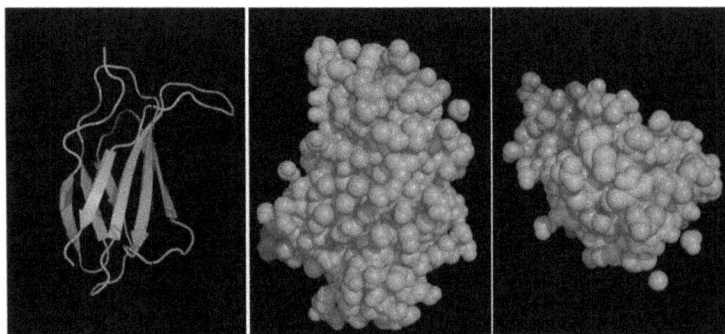

Figure 10.5. The FN-III scaffold used to generate the antibody mimic. The left panel shows the ribbon diagram of the tenth FN-III domain of FN (1fnf). The BC loop is colored magenta and the FG loop is colored red (RGD) and cyan (PAS, which is the segment that was deleted). The two right panels show space-filling models of another FN-III domain (pdb 1ten) with the BC loop magenta and the FG loop red. The five colored amino acids of each loop were randomly mutated to generate binding affinity [7]. Displayed with *PyMOL*.

termed 'Nano bodies' [6], can be selected for high affinity binding to almost any antigen.

One does not even need an Ig domain to make an antibody mimic. Koide and colleagues decided to explore an FN-III domain as an antibody mimic [7]. The FN-III domain has a beta sandwich structure very similar to that of the Ig domain. A particular advantage of the FN-III domain is that it has no disulfides, and can be produced in large quantity by expression in bacteria (disulfides cause problems for expression in bacteria).

Koide *et al* chose the tenth FN-III domain of FN, which they named FN3, for their scaffold. This domain had been characterized by crystallography and NMR. It is also extremely stable, remaining folded at 100 °C. A disadvantage of domain 10 is that its FG loop with the RGD is extended by three amino acids and protrudes as a flexible structure. Koide's group removed three amino acids to eliminate this flexibility. They decided to keep the RGD and delete the adjacent PAS. This should leave the FG loop as a shorter, rigid structure comparable to other FN-III domains (figure 10.5). The authors noted that the BC and FG loops were in a similar configuration to the CDR1 and CDR3 loops of the VH Ig domain, and formed a contiguous surface that might be modified to make a binding site. To generate the binding site they randomly mutated the five amino acids in each loop using site-directed mutagenesis with primers degenerate at the designated sites.

As a target antigen Koide *et al* chose the protein ubiquitin, to which FN3 had no binding affinity. They used the phage display technique to select for mutants that had affinity for ubiquitin. Phage display is a powerful sorting technique that has been used in thousands of studies. George P Smith discovered that he could insert short peptides or even whole protein domains into the coat protein pIII of the filamentous phage M13, and the insert did not compromise viability of the phage [8]. The insert is exposed on the surface where it can bind an antibody or any other binding partner. In the Koide application the library of randomly mutated FN3 domains was inserted into pIII, and the phage were selected for ubiquitin binding by incubating them in plastic wells coated with ubiquitin. The wells were washed to remove unbound phage, and bound phage were eluted with soluble ubiquitin. This panning was repeated five times, and selected phage were sequenced. The table shows three partial protein sequences from the enriched pool. One clone, dubbed Ubi4, dominated the pool and was selected for further study.

Name	BC loop	FG loop
Wild type	AVTVR	GRGDS
Cl1 (Ubi4)	SRLRR	PPWRV
Cl2	GQRTF	RRWWA
Cl3	ARWTL	RRWWW

The authors first used a simple ELISA to confirm the binding, then followed with a more quantitative competition ELISA (described in chapter 9) to determine 'the concentration of the free ligand which causes 50% inhibition of binding … is approximately 5 μM.' This is a good estimate of the K_D.

Koide *et al* [7] then expressed the Ubi4-FN3 protein in *E. coli*, and noted that it was much less soluble than the wild type FN3. They substantially improved the solubility 'by adding a solubility tail, GKKGK, as a C-terminal extension' and named this protein Ubi4-K. They measured its folding stability by denaturation in GuHCl. The ΔG_O for unfolding at neutral pH was 7.2 kcal mol^{-1} for wild type FN3, and 4.8 kcal mol^{-1} for Ubi4-K. This means that Ubi4-K was destabilized relative to wild type, but since FN3 is such a stable domain, the Ubi4-K retained a healthy stability. Comparison of Ubi4-K and the original FN3 by NMR showed 'Only small differences … in the chemical shifts, except for those in and near the mutated BC and FG loops.'

An important point in this initial study was that their library of random mutants contained only ~10^8 independent clones, which was a tiny fraction of the 10^{13} possible sequences from ten randomized amino acids. One could not expect to find the highest affinity binders in this limited library. Subsequent refinements have improved the affinity from 5 μM to the nM range, by additional rounds of mutation of the same ten amino acids.

The original Koide study developed FN3 to acquire a binding activity for ubiquitin, but the method is actually universal. The FN3 scaffold has now been used to develop binding reagents to hundreds of different antigens, with typical affinities in the nM range. Some of these have reached clinical applications [9].

FN3 is not the only scaffold for antibody mimics. The FN3 domain exploited by Koide *et al* [7] had several advantages over previous scaffolds, but a variety of scaffolds with unique advantages has now been developed. There is a rich literature, which one can access in recent reviews [10, 11]. *Methods in Enzymology* vol 503 (2012) has chapters covering several of the most important antibody mimic scaffolds.

References

[1] Pierschbacher M D and Ruoslahti E 1984 Cell attachment activity of fibronectin can be duplicated by small synthetic fragments of the molecule *Nature* **309** 30–3

[2] Leahy D J, Aukhil I and Erickson H P 1996 2.0 Å crystal structure of a four-domain segment of human fibronectin encompassing the RGD loop and synergy region *Cell* **84** 155–64

[3] Van Agthoven J F, Xiong J P, Alonso J L, Rui X, Adair B D, Goodman S L and Arnaout M A 2014 Structural basis for pure antagonism of integrin alphaVbeta3 by a high-affinity form of fibronectin *Nat. Struct. Mol. Biol.* **21** 383–88

[4] Amit A G, Mariuzza R A, Phillips S E and Poljak R J 1986 Three-dimensional structure of an antigen–antibody complex at 2.8 A resolution *Science* **233** 747–53

[5] Frank R 2002 The SPOT-synthesis technique. Synthetic peptide arrays on membrane supports—principles and applications *J. Immunol. Methods* **267** 13–26

[6] Pardon E, Laeremans T, Triest S, Rasmussen S G, Wohlkonig A, Ruf A, Muyldermans S, Hol W J G, Kobilka B K and Steyaert J 2014 A general protocol for the generation of nanobodies for structural biology *Nat. Protoc.* **9** 674–93

[7] Koide A, Bailey C W, Huang X and Koide S 1998 The fibronectin type III domain as a scaffold for novel binding proteins *J. Mol. Biol.* **284** 1141–51

[8] Smith G P 1985 Filamentous fusion phage: novel expression vectors that display cloned antigens on the virion surface *Science* **228** 1315–17

[9] Bloom L and Calabro V 2009 FN3: a new protein scaffold reaches the clinic *Drug Discovery Today*

[10] Jost C and Pluckthun A 2014 Engineered proteins with desired specificity: DARPins, other alternative scaffolds and bispecific IgGs *Curr. Opin. Struct. Biol.* **27** 102–12

[11] Gilbreth R N and Koide S 2012 Structural insights for engineering binding proteins based on non-antibody scaffolds *Curr. Opin. Struct. Biol.* **22** 413–20

IOP Publishing

Principles of Protein–Protein Association

Harold P Erickson

Chapter 11

Association of intrinsically disordered proteins—flexible binding partners

Directed reading:

Oldfield C J, Meng J, Yang J Y, Yang M Q, Uversky V N and Dunker A K 2008 Flexible nets: disorder and induced fit in the associations of p53 and 14–3–3 with their partners *BMC Genomics* 9 Suppl 1:S1 [1].

The previous chapters discussed the association of globular proteins to each other, a topic dominated by the lock and key paradigm of Chothia and Janin. In the past twenty or so years, it has been realized that a substantial fraction of proteins lack a rigid globular structure or have a substantial segment of peptide lacking rigid structure. These are now classified as Intrinsically Disordered Proteins/Peptides (IDPs). The disordered peptide is flexible and passes through a wide range of conformations, similar to chemically denatured proteins. IDPs are not suitable for enzymatic catalysis, but they participate in many protein–protein associations. In fact, their flexibility seems to suit them to binding multiple partners.

We generally like to use important research papers for course reading, but we would need several papers to cover the variety of associations of IDPs. Fortunately the 2008 review of Oldfield *et al* [1] does a great job of covering this field, using two prototypical examples: proteins named p53 and 14–3–3ζ. This is a long article, and the following sections may be considered optional (they are primarily concerned with induced fit motions in the globular domain, which we discussed in chapter 4).

Optional sections for directed reading of Oldfield *et al* are:

- Analysis of associations involving p53 using 3D structures.
- 14–3–3 binding to two different partners
- Figures 2, 5, 7.9

Their diagram of the protein p53, shown here in figure 11.1, is complicated and rich in details that will be explained. Start at the bottom.

doi:10.1088/2053-2563/ab19bach11

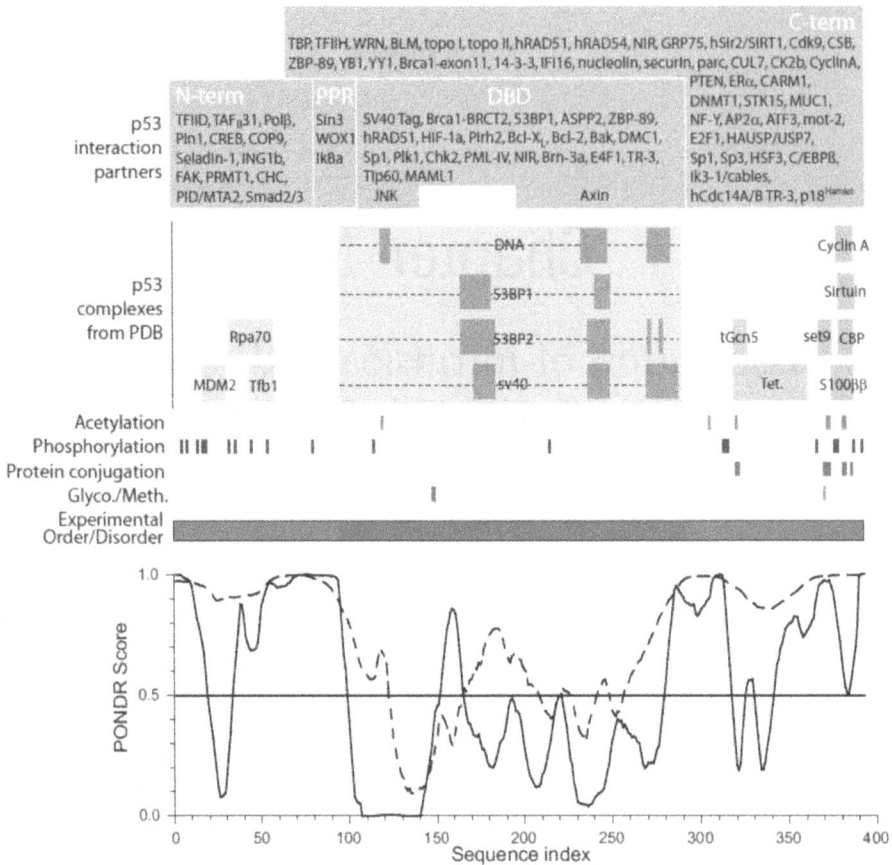

Figure 11.1. Diagram of protein p53 showing ordered and disordered segments and binding partners. See text for discussion. Reprinted from [1] (open access). Copyright Oldfield *et al*; licensee BioMed Central Ltd. 2008 CC. BY 2.0.

- The bottom graph shows the PONDR score, which is an evaluation of the likelihood for disorder that is based solely on the sequence. The *x*-axis is the sequence, and everything in the figure is scaled to it. The web site pondr.com allows the user to enter any protein sequence for evaluation of a PONDR score. The VLXT algorithm (dashed curve) is thought to have the highest accuracy while VSL2 (solid curve) is better at finding potential binding regions within longer disordered segments. A PONDR score greater than 0.5 indicates disorder. From this graph, one concludes that the N- and C-terminal segments are disordered and the central domain is ordered. This is indicated on the bar above the graph, where red indicates disordered and blue the ordered segments.
- Lines above the bar indicate the locations of post-translational modifications, phosphorylation, acetylation, etc. These are much more frequent in the disordered segments than the central globular domain.

- The top two panels show 'p53 interaction partners' and 'p53 complexes from PDB.' The general list of interaction partners is impressive for the number, but the important information is in the panel showing binding complexes with a PDB crystal structure.

Look first at the central globular domain, which does not involve IDPs. It binds three different proteins and DNA, each of them contacting 2–3 separate small segments of sequence (the contact peptides are indicated by blue bars). This is similar to the case of the antibody D1.3- HEL complex, where the antigenic site comprised two separate segments that were separated in the sequence but clustered together in the 3D structure ([2] and chapter 4). Also note that the three different proteins all bind the same patch on the p53 globular domain. This is similar to the case of GHR, where the same patch on GHR binds two different patches on opposite sides of GH ([3] and chapter 5).

Now we turn our attention to the C-terminal IDP, especially the four binding partners Cyclin A, Sirtuin, CBP and S100ββ. These are globular proteins, and, as indicated in figure 11.1, they all bind the same short sequence of the C-terminal p53 IDP. Figure 11.2 shows the crystal structures of these four complexes. The same short segments of 7–12 amino acids from p53 adopt different structures when binding the different globular proteins: alpha helix, beta strand and two coil structures. Note that this IDP is flexible and can adopt multiple conformations before binding one of the globular proteins. After binding it is locked into a rigid structure, but this is different for the each globular binding partner.

The paper then turns to 14–3–3ζ, a globular protein that binds a number of different IDPs. Crystal structures are available for at least five different complexes.

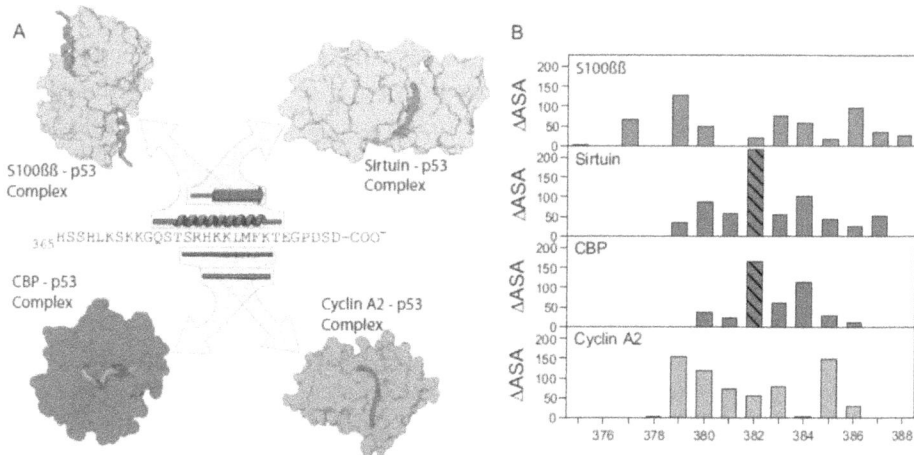

Figure 11.2. (a) Crystal structures of the Ct peptide of p53 binding to four different globular proteins. In the top two structures the peptide adopts an alpha helix and a beta strand. In the bottom two it adopts a different coil structure in each. (b) The WASA buried by each amino acid of the peptide in each structure. The hatched bars are acetylated lysine. Reproduced from [1] (open access). Copyright Oldfield *et al*; licensee BioMed Central Ltd. 2008 CC. BY 2.0.

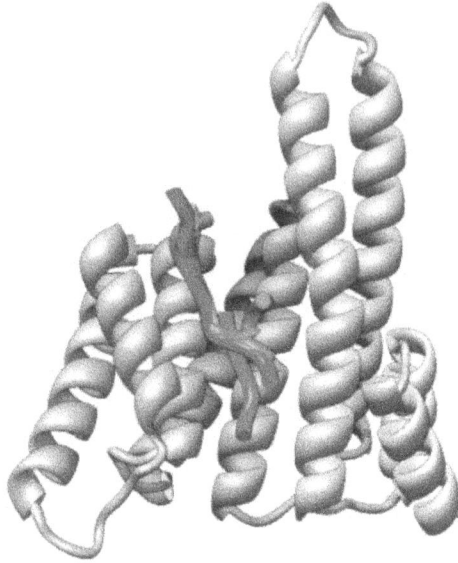

Figure 11.3. The structure of 14–3–3ζ showing five bound peptides from different IDP binding partners. Reprinted from [1] (open access). Copyright Oldfield *et al*; licensee BioMed Central Ltd. 2008 CC. BY 2.0.

Figure 11.3 shows the structures of the ligand peptides on the common structure of 14–3–3ζ. The bound peptides are short, only 5–8 amino acids, and they show partial sequence identity. Importantly, all five peptides bind in the same groove on 14–3–3ζ. They are almost superimposed near the middle, where most have a phospho-Ser/Tyr, but diverge somewhat toward the ends. The peptide backbone of 14–3–3ζ showed minimal changes upon binding the different peptides, although there were some side chain movements to accommodate the different peptides.

The IDPs appear to form stable bonds with a much smaller buried surface area than globular proteins. The interface of two globular proteins buries ~600–800 Å^2 WASA for each side of the interface (chapter 3). (This area would apply approximately to the hot spot for an inefficient bond like GH–GHR (chapter 5)). The total area buried by the short IDP is less than half this, as tabulated in figure 11.2(b). This suggests that the bond holding an IDP to its globular protein partner is strengthened by contributions other than buried WASA. One candidate is the multiple H bonds that the IDPs make to the globular protein partner (see figure 3.3 in chapter 3 for the H-bonds formed by the m1 peptide binding to 14–3–3ζ).

The two examples above showed the binding of an IDP to a globular protein partner, where the bound IDP adopted a defined structure. It is also possible for IDPs to bind other IDPs. One example is the 30 amino-acid tetramerization domain of p53. This is intrinsically disordered as a monomer, but it acquires a defined structure of a beta strand and alpha helix upon association. The structural features of the tetramer and the consequences of tetramerization for function are reviewed in [4]. A more surprising association of two IDPs is that of H1 and ProTα. H1 is a ~200 amino acid protein, mostly disordered and highly basic. ProTa is a ~110 amino-acid

IDP and highly acidic. These two IDPs form an ultrahigh-affinity complex in which the peptides retain their disordered state and motion [5]. The dimers are apparently held together by electrostatic interaction of the oppositely charged peptides. Finally, we should note that IDPs play key roles in the formation of biomolecular condensates, cytoplasmic organelles formed by liquid–liquid phase separation [6].

A general conclusion is that IDPs participate in a wide variety of protein associations. A single short IDP can bind to several different globular proteins, and a given globular protein can bind several different IDPs. The binding of multiple partners is occasionally seen when both are globular domains. When one partner is an IDP this multiplicity seems more common.

The field of IDPs is growing in many directions. A review by Wright and Dyson [7] discusses complexities and mechanisms beyond the presentation here.

References

[1] Oldfield C J, Meng J, Yang J Y, Yang M Q, Uversky V N and Dunker A K 2008 Flexible nets: disorder and induced fit in the associations of p53 and 14-3-3 with their partners *BMC Genom.* **9** Suppl 1:S1

[2] Amit A G, Mariuzza R A, Phillips S E and Poljak R J 1986 Three-dimensional structure of an antigen–antibody complex at 2.8 A resolution *Science* **233** 747–53

[3] Cunningham B C, Ultsch M, DeVos A M, Mulkerrin M G, Clauser K R and Wells J A 1991 Dimerization of the extracellular domain of the human growth hormone receptor by a single hormone molecule *Science* **254** 821–5

[4] Kamada R, Toguchi Y, Nomura T, Imagawa T and Sakaguchi K 2016 Tetramer formation of tumor suppressor protein p53: structure, function, and applications *Biopolymers* **106** 598–612

[5] Borgia A *et al* 2018 Extreme disorder in an ultrahigh-affinity protein complex *Nature* **555** 61–6

[6] Ditlev J A, Case L B and Rosen M K 2018 Who's in and who's out-compositional control of biomolecular condensates *J. Mol. Biol.* **430** 4666–84

[7] Wright P E and Dyson H J 2015 Intrinsically disordered proteins in cellular signalling and regulation: *Nat. Rev. Mol. Cell Biol.* **16** 18–29

www.ingramcontent.com/pod-product-compliance
Lightning Source LLC
Chambersburg PA
CBHW082109210326
41599CB00033B/6640